Translation of Light

My Scientist Father, Warren Weaver And Our Conversations About Astrology

Helen Weaver

Translation of Light

Sally Weaver

Published by ACS Publications, Starcrafts LLC

Copyright © 2024 by the Michelsen-Simms Family Trust

All rights reserved. No part of this book may be reproduced or used in any form or by any means—graphic, electronic or mechanical, including photocopying, mimeographing, recording, taping or information storage and retrieval systems— without written permission from the publisher. A reviewer may quote brief passage.

First Printing 2024

Layout design by Sally Weaver
Cover design by Molly Sullivan

ISBN: 978-1-934976-70-8

Printed in the United States of America

for my father, Warren Weaver

a scientist who listened

Woodstock, New York
2020

CONTENTS

Introduction: i

Prologue: A Parking Lot for Chariots: 1

Chapter 1: Early Years: 7

Chapter 2: Daddy's Girl: 16

Chapter 3: World War II and Computers: 24

Chapter 4: A Woman of Letters: 32

Chapter 5: Probability and ESP: 41

Chapter 6: Rebellion: 48

Chapter 7: Science and Religion 59

Chapter 8: The Light of Greece: 67

Chapter 9: A Mad Tea Party: 73

Chapter 10: Dad's Humor: 79

Chapter 11: Why Astrology? 83

Chapter 12: Learning a New Language: 93

Chapter 13: Dialogue: 102

Chapter 14: Woodstock: 112

Chapter 15: Riverby: 119

Chapter 16: Astrology on Trial: 125

Chapter 17: The Dimensions of Personality: 133

Chapter 18: Warren Weaver on Modern Physics: 141

Chapter 19: The Unity of the Universe: 147

Chapter 20: Warren Weaver on Astrology: 152

Chapter 21: Changes: 158

Chapter 22: Lacrimae Rerum: 166

Appendix No. 1: The Weavers and their Stars: 176

Appendix No. 2: A History of Astrology: 195

Notes: 268

Bibliography: 283

Acknowledgments: 268

INTRODUCTION

This book is actually two books.

In the first place, it's a book about my father, the late Warren Weaver, a distinguished scientist and mathematician - so distinguished, in fact, that New York University named their math building Warren Weaver Hall. Vice president and director of natural sciences at the Rockefeller Foundation, world traveler, author of seven books and hundreds of articles, and a brilliant, kind, and charming man, he was my hero. This is a book about my father and me. So I guess you could call it a memoir.

But it's more than that. Sometime in the sixties my father became mildly alarmed when I became a serious student of astrology. He was concerned that his intelligent, well-educated daughter, as he believed me to be, could wish to devote her time to such an activity.

For years, we couldn't talk about it at all. Eventually we began discussing it politely in letters and finally, toward the end of his life, when I was interviewing him on tape about his life and travels, he became intrigued by what he called the astrological hypothesis, and we found ourselves engaged in a dialogue about astrology and science. We planned to write a book, to be called *Dimensions of Personality* (his title). But shortly after we began work on that project his health took a turn for the worse, and when Knopf offered me a major translating job, he urged me to take it. When I received the manuscript - it was 1200 pages - I knew that the book that Dad and I had planned to write was not to be. He died in 1978.

Nevertheless this is, secondly, a book about astrology. It's not the book that Dad had in mind - that book was made impossible by his death - but it's as close to that book as I can get. For example, Dad wanted our book to be in dialogue form - a traditional form of exposition in the history of Western philosophy - so I have incorporated some of our conversations that made it onto tape. He thought our book should include a history of astrology, for he had always felt that something with a long history should not be disregarded. So with the help of some modern astrologers and academics who have specialized in the subject, I have put together a little history of astrology. And since at 72 pages it is too long to go in the so-called memoir, I have relegated it to an appendix. Also in an appendix is a discussion of the birth charts of the Weaver family which seemed too technical for the non-astrologer.

If this is a book about my father, why is there so much about me? Because I want to show how each of us developed a philosophy that made possible such an unusual meeting of minds. And when I discuss the birth charts of the Weavers, the reader needs to know enough about both Dad and me to see how astrology works.

Because it does work! For as the great astronomer Johannes Kepler, who discovered the three laws of planetary motion, wrote, "The belief in the effect of the constellations derives in the first place from experience, which is so convincing that it can be denied only by those who have not examined it."

Astrology, the study of the stars, is at least 7,000 years old: almost as old, if not older than, recorded history. For all but the last three centuries of that time, astrology has been closely associated with astronomy. In ancient Babylonia and Egypt observation of the heavenly bodies was the province of a priestly class who also engaged in the prediction of terrestrial events.

The medieval mind was dominated by the idea that each individual was a microcosm, a miniature version of the cosmos, or macrocosm, and that everything in nature was linked together by a system of mysterious but discoverable sympathies and correspondences.

Although the Christian church made a few sporadic efforts to counteract belief in astrology, the notion that the lives of people on Earth are connected in some way with events in the heavens remained deeply rooted in the human psyche on every level of the social hierarchy. While not all astrologers were astronomers, a great many were, and some of the greatest astronomers the world has known, including Ptolemy, Tycho Brahe, Johannes Kepler, and Galileo, were lifelong practicing astrologers.

This situation of peaceful coexistence between astrology and astronomy continued well into the seventeenth century when, for a variety of reasons, astrology, once known as the "queen of sciences," fell into disfavor, first with scientists, and then with educated people in general. Partly because of a growing antipathy to the determinism that pervaded most popular astrology, and partly because astrology was assumed to depend upon a conception of the universe that had been made obsolete by recent advances in science, astrology went into a decline from which it is only beginning to recover.

Whereas in the Renaissance it had been an integral part of the curriculum of every well-educated gentleman, astrology was henceforth banished from the universities. The result was that although astrological symbols continue to exert a fascination over the minds of ordinary people, until quite recently few really first-rate minds have been drawn to the serious study of astrology. Cut off from the mainstream of scientific activity, a field of inquiry that once engaged the respectful attention of thinkers like Pythagoras, Plato, Plotinus, Thomas Aquinas, Roger Bacon, Martin Luther, Francis Bacon, John Dryden, and Goethe, gradually declined into a rather anemic vestige of its former vigor.

Until a few years ago astrology survived chiefly in an oversimplified form (that bears little or no relation to serious astrology) in the syndicated columns of daily newspapers and on the lunatic fringes of society, as a quaint pastime for a few harmless oddballs. Now, all that is changing. After three centuries of dormancy, astrology is in the throes of rebirth. The underlying reasons for this resurgence are as complex as the causes of its decline. One contributing factor may be the decline in this century of the influence of organized religion. And the relatively new sciences of psychology and psychiatry, aimed at helping people live more constructive lives through greater self-understanding, often tend to ignore the spiritual dimension, our need to feel some sense of connection with the cosmic framework of our lives.

The sciences in general seem to have gone too far in the direction of technology for its own sake. The occult renaissance of the 1960s - the sudden popularity of long-ignored subjects like astrology, palmistry, herbal medicine, acupuncture, Tarot, *I Ching*, Zen Buddhism, reincarnation, ESP, and so on - might be compared to the return of the repressed in modern psychology. An unconscious collective impulse to restore a balance has been violated by over-development of the rational mind at the expense of the capacity for belief and wonder, and an overemphasis on controlling our environment at the expense of trying to understand it and live in harmony with it.

The generation born during and after World War II tends to be more tolerant toward the possibility of planetary influence on the human psyche. Encouraged by the example of the great modern psychologist Jung, who openly admired astrology and used the horoscope as a diagnostic tool in his clinical work, a significant number of the members of this new generation regard astrology as a serious field of study. Serious books about astrology are being published in increasing numbers. In response to a growing demand, courses in astrology are once again being offered at the university level for the first time since the Renaissance.

Computers have entered the picture. Originally enlisted in the interest of commercial exploitation, to produce quickie charts for the masses (e.g., Grand Central's Astroflash), they are now being used in statistical studies designed to test the validity of astrological assumptions.

All this seems to suggest that after three centuries of banishment, astrology is about to come once again into its own and be integrated into the age of technology. But in this picture of increasing visibility one very important ingredient is lacking: scientific recognition, and the true social respectability that such recognition would confer. As long as scientists continue to dismiss astrology as superstitious nonsense, astrologers will be unable to get funding to undertake the research necessary to purge astrology of irrelevant archaisms and bring it up to date.

Scientists tend to think of astrologers at best as well-meaning but deluded eccentrics; at worst, as unscrupulous charlatans. As for astrologers, after being looked down on for so many generations, they tend to be uncomfortable and defensive in the presence of scientists. The two groups tend to avoid direct contact, contenting themselves with retiring into their professional enclaves and muttering behind each other's backs. When scientists bother to notice astrologers at all, it is to express alarm at the proliferation of unreason they help to promote and contempt at their claim that the nonsense they perpetuate is also a science.

This situation is just beginning to change. By the turn of the twenty-first century, reputable scientists had started speaking at astrological conferences: a statistician here, a meteorologist there, a biologist here, a physicist there, had come forward and at the risk of jeopardizing his or her professional reputation, had admitted to observing phenomena that offer compelling support for astrology. A small but growing number of physicians and psychiatrists, following in Jung's footsteps, are working with astrologers on the diagnosis and treatment of mental and physical problems.

Encouraged by the growing body of evidence for their claims, astrologers have started coming out of their ghettos and attempting to make contact with members of the scientific establishment. The long overdue cross-fertilization has begun.

Yet, the fact remains that it's still something of a rarity for a scientist and an astrologer to be able to sit down and talk about the nature of the universe in a relaxed and friendly way. Some of my father's questions took several years to answer; some of the most interesting ones remain unanswered still. But somewhere along the line, he became intrigued by the astrological hypothesis: the idea that significant correlations exist between the positions of the Sun, Moon, and planets at birth and personality traits of human beings on Earth. Somewhere along the line, he decided that this hypothesis *was* worth investigating. If there wasn't anything to it, why had so many millions of people believed in it for so many centuries? And if there *was* anything to it, shouldn't scientists, who now regard human behavior as a valid field of study, be exploring the possibilities that it raised?

It was once felt to be inconceivable that the Earth was round, or that the Sun was the center of the solar system, or that human beings had evolved from lower forms of life, or that much of human behavior is influenced by motivations of which we are not entirely conscious; yet all of these "inconceivable" ideas are now generally accepted by scientists.

I know that I speak for my father when I add these lines from the first draft of our book, written back in 1976 when he and I were working on it together:

With all due humility, my father and I feel that our dialogue is a rather special one. We have found it stimulating and rewarding to share our ideas in this way, and each of us feels greatly enriched by the experience. Among the many rewards it has brought are the discovery of ideas that science and astrology hold in common - bridge ideas, you might call them - and a renewed respect for the scientific method, which can and should be applied to any body of data, no matter how far-fetched it might appear at first glance.

Our purpose in writing this book has simply been to put our heads together in an effort to find a common meeting ground for astrology and science. In a sense, the simple decision to communicate creates that ground. We have tried to define a space where other astrologers and scientists could meet and exchange ideas. We feel that communication is a good thing in and of itself and that no person who is seriously trying to understand the nature of the world around them should have to work in a vacuum, or in an atmosphere of ridicule and scorn.

In a little piece she wrote on astrology for *Redbook* magazine, the anthropologist Margaret Mead ended by saying that she didn't know whether there was anything to astrology or not, but that the scientific attitude is always to keep an open mind. Whether our readers are scientists or astrologers or neither, we ask no more of them than that.

Helen Weaver and her father, Warren Weaver circa 1976

Prologue
A PARKING LOT FOR CHARIOTS

It's a poor sort of memory that only works backwards.

—The White Queen, *Through the Looking Glass*

"Are you warm enough, Dad?"

My father had on the old tan sleeveless sweater Mother had knitted him many Christmases ago. It had a few holes in it, but was still a faithful friend.

"I'm fine." I tucked Cheech's afghan around his legs for good measure.

"Ready?"

"Ready. Let's go!"

I poked "Record" on the little Radio Shack tape machine on the coffee table between us, and cleared my throat: "It's April twenty-fourth, nineteen seventy six. . .."

When my father was in his early eighties, his health had started to fail. It had really started to fail much earlier than that, but I guess I was too busy with my own life to notice, or maybe I was in denial. I remember that when I did start to notice that he was getting old, I wanted to shake him and say, "Stop pretending to be an old man!" I thought of it as "voluntary senility." (Exactly what my aides are accusing me of these days.) I was such a wise-ass in those days. But I also loved him, and didn't want to lose him, ever.

Compulsory retirement is a terrible thing. My father had to retire from his job as vice president of Rockefeller Foundation at sixty-five, even though he would have had many good years left in him. His friend Margaret Mead wrote that retirement is harder on men than it is on women because they're like an admiral without a fleet.[1] It was becoming clear to me that Dad was depressed. Inactivity was breeding a host of health problems. He needed a project.

By the 1970s I had to face the fact that he wasn't going to be around forever, and I wanted to hear his stories. Stories I had tuned out as a child because I resented his monologues at the dinner table, I now wanted to know.

In the spring of 1976 I was visiting my parents at their home in New Milford, Connecticut. I asked my father if he would answer a few questions on tape, and he liked the idea. He was used to being interviewed.

I began by bringing him various objects he had collected in his travels: a set of temple carvings from Kyoto; a nineteenth-century wooden stirrup from Mexico; a pre-Columbian wine vessel in the shape of an otter from Chile; his seashell collection exquisitely arranged in a rosewood cabinet from Hong Kong; a set of Egyptian prayer vials in a little brown box. And he would start to talk.

This morning it was the prayer vials: tiny glass things, obviously very old; the label on the bottom of the box said "Hotel Ramat-Aviv, Tel-Aviv, Israel."

"Where did you get these, Dad?"

"They were given to me by the man who was at that time the prime minister of Israel, Yigael Yadin."

"How did you happen to meet him?"

"Well, he was an archaeologist and a soldier. He was the head of the Israeli army when they won their short war, but a great archaeologist, and he entertained me in his home. In his home he had a sort of a display case with a lot of very, very old Egyptian glass that he was terribly interested in. He took me up on the heights that they've been having such a time fighting about – what do you call that mount north of Jerusalem?"

"Are they fighting about it now?"

"No, not now, but they were for a long time. The Golan Heights! He took me up there and showed me where they were making their diggings. And there was a big field we could look down on, and he said, 'It'll sound sort of ridiculous when I tell you this, but that was a parking lot. It was a parking lot for chariots.'

"He said, 'It sounds like a sort of a mixture of two different ages of time, but that's exactly what it was,' he said. 'It was a parking lot for chariots!'"

"They had to have parking lots too!"

"And while he was sitting there – he had one of these British canes that you stick in the ground and then can sit on the upper end of – you've seen those, haven't you?"

"Sure."

"He was sort of resting on this, and"–

"Was he an old fellow?"

"No, at that time I think he was probably in his thirties. Very attractive. And as we were sitting there and he was telling me about this parking lot for chariots, one of his workers rushed up to him and said to him something in what I assumed to be Hebrew. And he said, 'Oh, this is interesting. They just found this and thought I'd like to show it to you. It's a scarab.'

"Whenever they take a wall down, they'll take a stone out, and then they'll take a whisk broom and brush all the stuff out into a pan, and then at night they'll put all of that stuff through different size screens so that any fragment, no matter how small, will be rescued, you see."

"Amazing. And can they do this with solid rock? Or aren't these walls made of rock?"

"The walls are made of rock, but this was the loose material in between the rocks. He said they'd find fragments of little vases."

"He was so young, and he was getting to do that work?"

"Yes. His father was one of the greatest archaeologists that's ever lived."

"He must have been a remarkable man."

"He was, he was a very remarkable man. And they would save all these fragments and frequently from a whole pile of fragments they'd be able to reassemble a little vase that was made out of clay or something like that.

"But as I say, as we were sitting there and he was talking to me about this parking lot for chariots, a fellow brought this perfectly lovely little thing up and he showed it to me and he said, 'That's a scarab.' And I said, 'Well, it isn't Hebrew,' and he said 'Oh, no, that's Egyptian.' 'Well,' I said, 'how'd it get here?' He said, 'I don't know how it got here.' He said, 'We probably will know some day.'

"But it was a beautiful little thing. Looked like a beetle, you know."

"That wasn't one of the things that he gave you, though? You don't have any scarabs lying around the house?"

"No, I don't. A couple of days after that, back in Jerusalem, he entertained me at dinner at his home, and as I say, he had a big kind of display case that was just full of pieces of old Egyptian glass. He showed me a lot of them and told me what they were. And he said, 'This is a tear vase, and this would be used when a man would bury his wife, for example. This would go into the grave with her.'"

"Would it contain his tears?"

"At least symbolically, containing his tears." Dad picked up one of the shards.

"And this is a piece of old, old Egyptian glass of some kind of a vase. Well, I was so interested in these various things, and the next day when he took me down to put me on the airplane –

"Can you determine what year this was, Dad?"

"Let's see how close I can come to that. . .."

"When was Israel created? One of us is as bad as the other!" (At lunch that day, Mother told us it was 1948.)

"That I don't know. What's the island that the Turks and the Greeks have been having" –

"Cyprus?"

"Cyprus! He was putting me on the plane to go to Cyprus."

"To do what?"

"We wanted . . . let's see, what in the world were we doing in Cyprus? We weren't going to do any agricultural work in Cyprus." (Dad used to travel widely in connection with the Rockefeller Foundation's agricultural projects.)

"For the life of me, I can't remember why we were going there. You see, you couldn't fly from east of there; you couldn't fly and land in Cyprus, you had to land over in Israel some place and then shift over to a different type of plane that was allowed to land in Cyprus.

"And he was taking me down to that plane, and just at the foot of the steps going up to the plane as he was shaking hands to say goodbye to me, he pressed a little box into my hand.

"I said, 'What's this?' He said, 'Don't look at it now, or you won't take it, but I want you to take it with you.' He had noticed that I had admired these two things and had picked them out and gave them to me that way."

And now I must sift through Dad's and my obsessive archives – the carefully captioned photo albums, the letters, the journals, and the tapes – and try to put together the pieces of the book that, much to our surprise, decided to come out of these conversations. The book that he and I wanted to write together, before he got so tired.

At eighty-four, I'm getting a little tired myself. But like the archaeologist who pores over the fragments of the past, I have to hope that the pieces will somehow form a whole.

Warren Weaver with his older brother Paul and his mother, Kitty Belle. (circa 1897)

Chapter One

EARLY YEARS

The greatest single thought that has come through science, to my mind, is that of the unity of this living universe. There is no such thing as inert matter. Every atom is living, is partaking of the life of the universe. [2]

Max Mason

Warren Weaver was born July 17, 1894 in Reedsburg, Wisconsin, a town whose population at that time was, according to my father, "something under two thousand persons." His father, Isaiah Weaver, owned a drug store and was lead tenor in the Presbyterian church choir. His ancestors were pious German Lutherans who thought nothing of walking miles to church on Sunday morning and again to prayer meeting on Wednesday evening. German was spoken in the home and English at school, so my father grew up bilingual.

Isaiah was the youngest son of Jesse Weaver, a shoe cobbler and farmer whose father was born on the ocean to parents who were weavers of cloth in Strasbourg. Jesse was born Josiah Weber in Wayne County, Ohio. His name was changed when his baptismal certificate was translated from the German.

Dad's mother, Kittie Belle Stupfell, was born in Sharon, Connecticut, the only child of Emma and John Stupfell. My father was particularly fond of his maternal grandfather, a remarkable man who learned to drive an automobile in his seventies and, at the time of his fiftieth wedding anniversary, wrote out his memoirs by hand.

In this document, which my father transcribed without correcting the spelling ("I inherited my talent for spelling from Grandpa") John Stupfell tells that at age twenty-two he met his future wife, Emma Clapp, then a girl of fifteen, when the two of them stayed by the bedside of his sister's dying child: "There was a Spark Started as a result of that meeting, which was later fanned into a flame, which has never been quenched these fifty four years."

Isaiah met Kittie Belle when he came to Sharon to take a course in pharmacy. They were married in 1887 and had two sons, Paul and Warren. Sadly, Kittie Belle died in 1906 when my father was twelve years old.

Not long after the death of his mother he entered the seventh grade and promptly fell in love with his teacher, Miss Cicely Kavanaugh. In his autobiography, *Scene of Change*, he writes:

Indeed, my reports home were so glowing that my lonesome widowed father decided to visit the seventh grade at Washington School to see how much I had exaggerated. He must have concluded that for once I was right, for midway in my high school years he married her.[3]

My father was devoted to his stepmother, who later became known in our family as Cheech. This nickname came about in the 1930s when I couldn't pronounce Cicely, so "Cheechee" was shortened to "Cheech."

In the first chapter of his autobiography Warren Weaver tells about a gift his father brought him from Chicago when the family was living in Reedsburg. It was a one-dollar Ajax electric motor and a dry cell to run it. Dad doesn't say how old he was, but it was clearly one of his earliest – and happiest – memories:

Within a week I had built from spools and such like all the little rotating devices which the tiny torque of the motor could manage. Then I began to penetrate its mysteries.

Dad describes his joy as he took the thing apart several different ways and put it back together, and wonder of wonders, it still ran! He decided then and there that he wanted to spend his life building motors. He had no idea what you call persons who did this, but some adult, sensing his enthusiasm, remarked that he would probably grow up to be an engineer.

Although it was years before he really understood what the word meant, he immediately adopted it as his calling. By the time he had to register for his college courses at the University of Wisconsin he realized that there were varieties of engineering. His father took him to see a local civil engineer who went to the same church as the Weaver family for advice.

He asked, "Do you want to spend your life working for some big company, or would you like to look forward eventually to being your own boss?" The answer to that was obvious, so he said,

"Then you should train to be a civil engineer." That settled that –

Ironically, my father did, in a way, spend his life working for "some big company" – if that word can be applied to a non-profit like the Rockefeller Foundation – and loved every minute of it.

In his sophomore year at the university my father was introduced to differential calculus, and it was deep calling to deep. He realized that what he really wanted to do was not so much engineering as science, and especially mathematics. He persisted in the five-year course toward the degree in civil engineering, which he eventually earned, but took as many mathematics and physics courses as possible along the way. One of these, a course in electrodynamic theory with the brilliant physicist Max Mason, became the turning point of his life.

A graduate *magna cum laude* of the University of Göttingen in Germany, which was in those days the mathematics center of the world and a charming and eccentric genius, Mason became my father's favorite teacher, mentor, and lifelong friend. Mason had written his doctoral dissertation under the famous German mathematician David Hilbert, and my father was fascinated to learn that Mason had arrived at the solution to his thesis problem in a dream.

After Dad graduated from the University of Wisconsin and after a brief stint teaching in Pasadena and a year of wartime service in Washington, D.C., it was Mason who was largely responsible for his returning to Madison to teach mathematics. It was Mason who suggested that they write a book together: *The Electromagnetic Field* was published by the University of Chicago in 1929. And it was Mason, in his later capacity of president of the Rockefeller Foundation, who was ultimately responsible for my father's being offered a job in far-off New York City.

Sometime during his junior year as a student in Madison, Warren Weaver met a shy, intelligent beauty named Mary Hemenway who shared his love of reading and nature (if not of mathematics) and fell in love. He wooed her for four years before she consented to marry him.

I remember reading in one of Dad's letters – Mother kept all of them – that he asked her to marry him ten times before he so much as touched her hand. This was probably an exaggeration, something my father was prone to, but only a slight one: my parents were both born in the nineteenth century and it is no exaggeration to say that they had decidedly Victorian ideas about relations between the sexes.

From other letters I learned that my mother put off her marriage for four years because of the fear that had been instilled in her regarding the mysteries of physical love. ("Would it be sordid and awful?" she had written their mutual friend Dot.) But by the summer of 1919 she finally agreed to set the date. She chose September 1, shortly after my father's twenty-fifth birthday on July 17, but three days before her own twenty-fifth on September 4.

Dad writes: "According to her discontinuous theory of ages she was twenty-four until the day when she became twenty-five, so she had the satisfaction of being 'one year younger' than me on our marriage day."

In spite of my mother's fears and uncertainties, there was never any doubt in my father's mind that this was it. When she graduated from the University of Wisconsin and he was staying on for his fifth year, and even though she hadn't yet said yes, he gave her a set of the eleventh edition of the *Encyclopedia Britannica* as a graduation present. Mother was going to Carlsbad, New Mexico to teach Latin and ancient history, and she needed resources. But Dad was thinking and planning well beyond that. He knew that the future Weaver family ought to have a good reference library, and the eleventh was the great edition.

This may have been the first example of a family tradition that came to be known as Skates for Grandma: a gift given because the giver wants it available for his or her own purposes.

My parents were happily married for fifty-nine years, and my father never quite got over the fact that this lovely creature agreed to marry him: a skinny little guy with big ears and glasses who was good in math and terrible in gym. Even when they were old and gray he would look at her, shake his head, and say, "How did I ever get you?"

They were married by a Presbyterian minister in Mother's home in Carlsbad and took the night train to California, where my father had a job as assistant professor of mathematics at Throop College (later to become California Institute of Technology). After a year in Pasadena they returned to Madison, where they lived the idyllic, if somewhat parochial, life of a college professor and his family for twelve years. My brother, Warren Weaver, Jr., was born in 1923 and I came along in 1931. My father quickly advanced from assistant professor to associate professor to chairman of the mathematics department at the University. My parents loved Madison and their life there, and when the opportunity came to uproot themselves and move east, they were not sure they could bear to leave it.

During his interview at the Rockefeller Foundation my father explained that although his training had been in the so-called physical sciences – the branches of natural science that study non-living systems, such as astronomy, chemistry, geology, and physics – he was convinced that the wave of the future was going to be in the biological sciences:

The idea that the time was ripe for a great new change in biology was substantiated by the fact that the physical sciences had by then elaborated a whole battery of analytical and experimental procedures capable of probing into nature with a fineness and quantitative precision that would tremendously supplement the previous tools of biology–one can almost say "the previous tool," since the optical microscope had furnished so large a proportion of the detailed evidence.

Although the range and power of instrumentation available in the thirties for experimentation in biology and medicine was nothing like what my father correctly foresaw it would become, enough was perceptible to convince him that biology was about to enter a new era. Scientists, especially in the fields of genetics, physiology, and biochemistry, were on the verge of unlocking the secrets of the living cell. He told the people at Rockefeller that he thought they should move in this direction, but that he did not feel qualified to lead such a program:

But as I was enthusiastically convinced of the importance of moving in that direction, and because I did have the necessary background in the physical sciences, they somewhat rashly, as it seemed to me, offered me the directorship of that division of the Rockefeller Foundation dealing with all aspects of science other than professionally medical.

What Dad doesn't mention in his autobiography but often freely acknowledged was that his early prescient interest in biology and his sense of its profound coming importance was sparked in large measure by my mother, who was in turn inspired by a great biology course she took at the university. (Alas, I never learned the professor's name.) Given that the Rockefeller Foundation, under my father's leadership, led the world in its support of the emerging field of molecular biology, in a sense Mary Hemenway influenced the future course of science in America – and indeed, in the world.

Meanwhile, she also helped my father to accept Rockefeller's "rash" offer. For when in the course of their weighing the pros and cons of accepting it he said, "After all, we must make up our minds," Mother informed him, "Of course, we *have* made up our minds." They started to pack.

The Weaver family moved to New York and my father assumed his duties as Director of Natural Sciences at the Rockefeller Foundation in February of 1932. The European office of the Foundation was located in Paris. Less than three months after Dad joined the New York office in Rockefeller Center he left for Europe, taking all of us along.

One of my earliest memories is of the terrifying blast that came out of the smoke stacks on the Statendam and, on the return crossing (I had learned to walk in a croquet court near our *pension* in Saint-Cloud) of running across the deck and smelling the glorious sea air. To this day, those wonderful old ocean liners are my favorite means of travel, and I mourn their passing.

The senior science officer in the Paris office of the foundation was an organic chemist on leave from Princeton named Lauder W. Jones. Leaving Mother, my brother, and me at the *Pension Sivrais*, Dad and he traveled together to university and research centers all over Europe, looking for promising young scientists to whom to offer fellowships.

"Jonesy," as he was known to his intimates, happened to be a devout gourmet who was familiar with all the best restaurants, their *maîtres d'hôtel* and their cellarmen. My father was fond of quoting something Jonesy used to say toward the end of an especially lavish meal: "The human stomach has a normal capacity of approximately two pints – but *fortunately, it is capable of expansion!*"

Dad was still fluent in German but his French was only so-so, and the gender of French nouns was particularly troublesome. I remember his telling how one of his colleagues in the Paris office had solved this problem by coming up with an article that was located so precisely halfway between "*le*" and "*la*" that not even a Frenchman could tell the difference.

All this was heady stuff for a mathematics professor from a Midwestern university, but it was also hard work. Looking back over his life when he was writing his autobiography in the seventies, my father decided that the most important thing he had ever done was the redirecting of the considerable resources of the Rockefeller Foundation towards experimental biology.

He writes that when he read James Watson's exciting account of the discovery of the structure of DNA in his book *The Double Helix* in 1968, he was struck by the fact that of the people who had played leading roles in this historic breakthrough, twenty-four out of twenty-seven had received grants from the Rockefeller Foundation. Furthermore, out of eighteen Nobel laureates who had been involved with molecular biology between 1954 and 1965, fifteen had received grants from the Foundation. And all this was *before* they were awarded the Nobel Prize – indeed, on an average of nineteen years in advance.

The Rockefeller Foundation has, I firmly believe, a solid and authoritative basis for taking satisfaction in the role it played in emphasizing, over a period of over a quarter of a century, the support of research in quantitative biology.

Dad is too modest to point out that up until 1959, all of this support was given on his watch.

When the Weaver family first arrived in New York we perched briefly in an apartment on upper Fifth Avenue. I still have an elegant little floor plan my father made of this apartment, which he must have found for us during his first solo trip east, and sent to Mother back in Madison for her approval.

Eventually my parents decided to look for a house in Scarsdale. If they chose that archetype of suburban affluence in Westchester County as the best place to raise their young family it was not out of any sort of social snobbery: back in Wisconsin, my father had acquired a healthy disdain for sororities and fraternities, and he always prided himself on the fact that the Weavers never joined the Scarsdale Golf Club. But at that time, partly because of its wealth, Scarsdale had the best public school system in the country. My parents were both teachers, and the education of their children was a top priority.

Our house in Scarsdale was full of books: the complete works of Scott and Dickens and

Trollope, Shakespeare's plays in little leather-bound Temple editions, and many books of poetry: Browning, Byron, Keats, Milton, Shelley, Wordsworth, three volumes of Dickinson, four of Frost, and five of Edna St. Vincent Millay, my father's favorite, and why not? For "Euclid alone hath looked on beauty bare."

Dad had a lifelong interest in art and a passion for collecting and he and Mother both had exquisite taste, so the house also had oriental rugs, Stickley furniture, and beautiful etchings. Back in Madison, Frank Lloyd Wright had gotten in trouble with the IRS, which had confiscated part of his collection and sold it at auction, enabling my father to purchase some lovely Japanese prints on his teacher's salary. So my brother and I grew up surrounded by beautiful things.

In those days before television my parents read aloud to each other and to us, a habit they continued into old age. The A. A. Milne characters – Winnie the Pooh, Christopher Robin, Piglet, Eeyore, Tigger, Owl, Kanga and Roo – and Mole, Rat, Badger, and Toad from *The Wind in the Willows* – were all part of the family. Dad's favorite childhood book – and indeed, his favorite book, period – was *Alice in Wonderland*, by his fellow mathematician, Lewis Carroll. At some point he became a collector of Carroll, and eventually his collection became the largest in America and the second largest in the world.

Somehow, I never warmed to Alice. She always struck me as too cold and logical. Maybe I was jealous, as if she were a clever sister with whom I had to compete for my father's affections. I seem to remember Candice Bergen saying she'd been jealous of Charlie McCarthy, her ventriloquist father Edgar Bergen's dummy. It's certainly possible to be jealous of a fictional character, and fairly early in my life it became clear that I was "Daddy's girl."

Passport Photos of Warren, Mary, Warren Jr. and Helen Weaver 1931

Chapter Two
DADDY'S GIRL

Two roads diverged in a yellow wood,
And sorry I could not travel both
And be one traveler...

Robert Frost, *The Road not Taken*

To get the full flavor of my childhood in Scarsdale in the thirties, you have to remember that both of my parents were born in the nineteenth century in small towns in the Midwest. Mother's home town was Junction City, Kansas, which was not much bigger than Reedsburg, Wisconsin.

My mother's family came from Puritan stock. Her father, Edward Hemenway, was a kind man who kept a country store, whose employees stole from him, and who had such bad asthma that the family moved to Carlsbad, New Mexico in the hope that the dry climate would be better for him. Her mother, Helen Clark Hemenway, like Dad's father, was a pillar of her church who sang solos in the choir. She was also a spell-binding Bible teacher, a member of the Daughters of the American Revolution and the Women's Christian Temperance Union, and a direct descendant of William Bradford, first governor of Plymouth Colony and a God-fearing Puritan who had come over on the Mayflower.

I was named after Grandma Hemenway and although she was a rather formidable presence (she had one glass eye, which gave her a stern expression reminiscent of the American Bald Eagle), I always found her fascinating. Mother once told me that early in my childhood Grandma Hemenway had told her, "I think God has something very important for Helen to do." I think this was by way of consoling my mother, for by all accounts, I was a handful.

My parents didn't quite know what to do with me. While it was not exactly true that they believed children "should be seen but not heard," they made jokes to this effect and probably secretly wished it were still the rule. In any case, they did belong to the old school about sparing the rod and spoiling the child. When I was perceived to be bad I was taken across Mother's or Daddy's knee and spanked, with either a wooden yardstick or Mother's hairbrush.

When I was really little, before I started kindergarten, I wasn't allowed to go beyond the two ends of our block on Brite Avenue. But Greenacres Elementary School and its playground, swings, and seesaws, was only a few blocks away, I had been there many times, and I felt I had outgrown that rule. One day my best friend Janet Marvin, who lived across the street and was a year older than me, bullied me into going to the playground with her, even though she knew I wasn't allowed.

So I was spanked for going beyond the block, and I remember thinking that my parents had really made a mistake, and losing respect for them.

My brother was eight years older than me. Dody, the German Catholic maid who came with us from Madison, was old too, at least in my eyes. I was living in a house where everyone was older than me and where consequently I felt I had no power.

Maybe that's why I invented Stone Dust.

While my father's colleagues were busy developing the first plastics, dreaming about computers, and unlocking the secrets of the atom, my friend Willie Kriegsman and I were working on a technological breakthrough of our own. Willie lived across the street from us. He was the only kid in my class at Greenacres who was for Roosevelt instead of Wilkie. All the other kids wore buttons that had the word "no" with three lines under it, which was short for "No Third Term." Willie was different, and I liked that.

Stone Dust was made by rubbing two stones together. Somehow I had conceived the notion that this substance was a powerful explosive. Maybe some phrase from one of my father's dinner table monologues had seeped into my subconscious. At any rate, while Dad had a team of mathematicians busy inventing better anti-aircraft guns, I had a team of neighborhood kids busy producing a deadly secret weapon. We stored the Stone Dust in glass jars borrowed from our mothers. When my workers got bored with their job, I encouraged them by painting lurid pictures of what would happen when we had enough of the stuff to blow up the neighborhood.

Our laboratory was Willie's back yard. One day after the other kids had gone home Willie kissed me, and we decided to get married. When I announced this plan to Mother, she told me I couldn't marry Willie. When I asked her why she said it was because Willie was a Jew.

I had heard the word, but I didn't know what it meant. The way Mother said it, it didn't sound good. When I pressed her for an explanation, she just said that it would not be appropriate. The next time I saw Willie, I told him I couldn't marry him. When he asked me why, I said, "Because you're a Jew," knowing I was making a mistake but hoping he would shed some light on the mystery. But he just got a funny look on his face and sort of backed away. After that we didn't make Stone Dust any more.

My mother's anti-Semitism was short-lived. As a scientist, my father was aware that many of the most brilliant scientists the world had known were Jewish. It was impossible to be a member of our family and hold onto such an antiquated idea.

My best friend Janet and I were fascinated by the differences between male and female anatomy. She had seen her brothers, and I had had one look apiece at my father and my brother in their "birthday suits," but in both cases it was a very short look, as they both glared at me and took cover as soon as possible. So Janet and I got Willie (he must have forgiven me) to agree to show us his if we showed him ours. Willie honored his part of the bargain, but it must have been a hard act to follow, because we took one look, giggled, and changed our minds.

This event took place on the swing in my back yard. As ill luck would have it, Dody caught the whole scene out her bedroom window and snitched to Mother, who promptly warmed my bottom with her hairbrush. Not, to be sure, for breaking my word to Willie, for which I might have deserved punishment, but for the crime of looking at an eight-year-old circumcised penis, for "playing dirty."

After I got too old to be spanked I got lectures, which were almost worse, because they lasted longer. My father, especially, would go on and on about how selfish I was, and how the world didn't revolve around me, and how my parents would always love me, but I would never have any friends if I didn't improve. I didn't take this too seriously, for I was almost certain that the world *did* revolve around me. Dad's lecture always ended in a long and claustrophobic hug which I did not enjoy but had to stand there and endure. There was something about his smell that was just too strong for me.

It was a funny thing about hugs. My mother thought they were silly, and made fun of certain women of her acquaintance who were "huggers" and "kissers." She was not physically demonstrative with me, or with anyone else, for that matter. She was affectionate only when I was sick, which was a lot of the time. At about age five I developed chronic allergic rhinitis, which meant a perpetually stuffed-up nose and frequent respiratory infections. Mother called it Hay Fever, and said I had inherited my Grandpa Hemenway's nose.

When it came to hugs, Dad was just the opposite of Mother. It was clear almost from the beginning that my father doted on me even though I was "ornery," a "smart aleck," and "talked back." And in spite of the boring lectures and the suffocating hugs it was clear almost from the beginning that he and I had a very special bond. Dad had been sickly, too, and, as he puts it in his autobiography, "shy, introspective, not vigorous, and unskilled in the active games of childhood." He and I were the emotional ones, the ones with "the artistic temperament," as Mother put it, while she and my brother were the practical, unflappable ones, who kept their feelings to themselves.

I remember the first time I realized that, all things considered, I really preferred my father. I had just walked into the kitchen. My mother was standing at the sink, washing dishes. It must have been Dody's day off.

"Guess who I like best, you or Daddy?"

Mother ignored the question, so I repeated it.

Pretending not to be interested in the answer, which she undoubtedly knew, Mother contented herself with one of her favorite sayings: "Comparisons are odious."

But whether she wanted to know the answer or not, I told her.

"Daddy!" I shouted triumphantly, and rushed down the back stairs to the basement to watch him develop film or putter around in his shop, which was so much more interesting than helping her with the dishes.

Down in the basement of our house on Brite Avenue, in one corner of the room with the twin sinks, the round old-fashioned washing machine, and the mangler, a scary machine for wringing the water out of the laundry, my father had a workbench, and over my father's workbench, everything had its place. Each wrench, each screwdriver, each pair of pliers hung on the wall between two nails over a painted light-brown-outline of that tool. And so that there would be no question where each tool belonged, my father had painted its silhouette on the wall behind it in a darker shade of brown. So that on the wall behind the real collection of three-dimensional tools there was a sort of shadow shop, and if a tool was taken down, its shadow called the real tool back to it and away from the temporary disorder of the workbench.

This shrine to organization imprinted itself early on my mind. As did his respect for good tools. "If you want to do a good job, you have to start with good tools," he told me, returning a pair of pliers to its place.

I loved to watch my father work, and he liked to have me stand there watching him. Often minutes would go by with no other sound than our breathing – mine, the heavy mouth breathing of my chronic hay fever, – his, the faint humming of the motor of his mind. He had made my brother's cradle and his first set of blocks, bookshelves for my room, and a stool for his study, its padded leather seat secured by neat rows of brass tacks.

Later, when I was a teenager, and my parents were about to celebrate their silver wedding anniversary, I watched him design and construct a hinged wooden box just large enough to hold two hundred and fifty silver dollars, a symbolic offering for his bride.

That box made a big impression on me. In the first place, two hundred and fifty dollars was a lot of money in the forties. But that was the least of it. Imagine a man fashioning a treasure chest, not for the princess he hoped to win, but for the woman he'd been living with for twenty five years. I remember wondering whether anyone would ever feel that way about me.

My father, the scientist, the mathematician, the thinker of abstract thoughts, liked to make things with his hands. He was a superb dry-wall mason, and my first job was helping him move rocks at our summer home for fifty cents an hour. He was a gifted photographer and calligrapher. The technical drawings he made in his student notebooks as an undergraduate at the University of Wisconsin are exquisite. Two little watercolor landscapes that he did at age fourteen are remarkably good. In short, if he had not been seduced by science, my father could have been an artist. It was the road not taken, to borrow a phrase from his favorite Frost poem.

I think I must have inherited at least one of those artist genes. I certainly inherited the urge to create. Drawing and painting: this was my first love. If I was not the best artist in my class at Greenacres – Terry Friedman and I vied for that title – I thought I was, and when I was nine a drawing of mine was included in an exhibition at the American Museum of Natural History.

When I told my father I wanted to be an artist when I grew up he said that artists didn't make any money. I could tell he thought that was important. Nevertheless, he bought me an easel and paints and the finest sable brushes and paid for art lessons at the County Center in White Plains. He kept all of my drawings in a big black portfolio that I found later in the attic.

I must have been about six when I began begging for a piano.

My father, in case you haven't noticed, was generous to a fault. After all, his job at the Rockefeller Foundation consisted of giving away money; and he simply loved giving presents, especially to his family. He always went overboard at Christmas and birthdays and in fact every holiday provided him with an excuse to give his wife and children little gifts.

On my seventh birthday I was blindfolded and led into the living room and seated at the keyboard of a Steinway baby grand. Dad's brother, my Uncle Paul, who was chairman of the music department at Cornell, had picked out this wonderful old instrument and had it reconditioned. My parents were a class act.

For the next eight years I took lessons from Lois von Haupt, a German lady who lived just a few blocks away. It was Mother who made sure I practiced every morning before breakfast.

The summer I was seven, my parents rented a cottage at Candlewood Lake Club in Brookfield, Connecticut. Later, they bought land and built a cottage of their own, but that first year we stayed in the club cottage, and my brother and I shared a bedroom for the first and only time in our lives. It had bunk beds and he, being eight years older and a boy, got the top bunk.

I remember lying sprawled on the lower berth, leaning half over the edge, writing a letter to my friend Nancy back in Scarsdale. At seven, I could write well enough, but I suffered from an embarrassing lack of material. So I made up a lie, my first attempt at fiction: "Nipper threw up."

(My brother's nickname, later shortened to "Nip," was inspired by a character called "The Little Nipper" in a play my parents had fancied as newlyweds.)

"I DID NOT!" yelled my brother from the top bunk where he had been observing my literary efforts – effectively nipping, as it were, my story-telling impulse in the bud.

I was caught red-handed, guilty not only of lying, but of using someone for the sake of a story. Thus, did I become aware at an early age of my lack of talent for fiction.

I had a burning desire to write, but nothing to write about, a dilemma that continued to plague me for years.

Dr. Warren Weaver 1940. Principal Investigator, Analog Conversion Project

Chapter Three
WORLD WAR II AND COMPUTERS

Science is motivated by curiosity, inspired by imagination, and based on faith. [4]

Warren Weaver

Sometime in the early forties my father's dinner table monologues began to feature something called "the NDRC," and a little later, "the OSRD." As a morose and dreamy pre-adolescent I pretty much tuned out the details of his day at the office. Years later, I learned what those letters stood for, and that my father had been involved in some important work for the government during World War II.

Dad's extensive travel during the thirties and his awareness of what he referred to as "the Jewish refugee problem" convinced him early on that "something evil was taking place in Germany." A hater of war, he nevertheless realized well in advance of Pearl Harbor that the United States would have to become involved, and that he wanted to find some way to be useful, as he put it. He knew that President Roosevelt had appointed a scientist named Vannevar Bush as head of the National Defense Research Committee. A year later, its name was changed to the Office of Scientific Research and Development, "this latter being a more accurate title, as it had become clear that the organization would be involved not only in research but also in the practical stages of construction and testing of pilot models."[5]

Dad wrote Bush offering his services, was accepted, and was put in charge of what was known then as "fire control," a military term referring to "all the devices and procedures used to assure that any projectile (a shell fired from an anti-aircraft gun, a bomb dropped from an airplane, or a torpedo launched against a ship) will in fact hit the desired target."

The fire control group was concerned with all of these kinds of devices, but especially, as it turned out, with anti-aircraft equipment. Since an anti-aircraft gun must hit a moving target, such equipment must be complex: it must anticipate where the target will be when the projectile reaches it; it must calculate the correct angle of the gun and when it should be fired; and it must account for various ballistic and environmental factors as well. And, since the guns are heavy, it needs to be powerful.

What was needed, in other words, was a built-in computer. But computers were hardly even in their infancy and the equipment the army was using had barely advanced since World War I.

The bombing of England began on July 10, 1940 – which, as my father noted with characteristic fascination with the improbable, happened to be the very same day he first met with Vannevar Bush about his new task. The upshot of that meeting was that Dad began to assemble a team to attack the problem. The original core group consisted of an engineer from MIT named Samuel H. Caldwell, a mathematician from Bell Telephone Laboratories named Thornton Carl Fry, and an inventor friend of Dad's named Ed Poitras, who designed the automatic control system for the 200-inch telescope on Mount Palomar.

Other engineers and technical assistants were added to the original team, and all of these talented individuals worked together in harmony and with a sense of purpose. My father often commented on the "tragic shame" that war often brings out the best in people.

The most important part of their job turned out to be improving the capacity of ground weapons to shoot down enemy planes. The mechanical gears and cams previously used were not fast or accurate enough to cope with the targets of World War II. But just when they realized that a new approach would be necessary, Bell Telephone Labs came up with the idea of an electrical gun director. In *Scene of Change,* his autobiography, Dad notes that the original idea for this novel device had occurred to one of their engineers in a dream.

Dad's group acted as a liaison between Bell Labs and the U.S. and British military. By September of 1941 the first model for this new equipment was being produced. It was tested in November and by February 1942 it was accepted by the army.

In the summer of 1942 a new agency, the Applied Mathematics Panel, was created to bring the expertise of probability and statistics to bear on all sorts of problems involving military strategy. As chief of the AMP, my father directed the work of over three hundred mathematicians and was in constant contact with representatives of all branches of the military.

On June 6, 1944 – D Day – when the Germans tried to bombard the allied troops landing on the beach in Normandy, the first three planes they sent were shot down with Dad's team's equipment. The success of those first computerized tracking devices set a new record for anti-aircraft fire.

On June 12, 1944 the first German "buzz bombs" – an early guided missile – began to rain down terror on London. By this time three different devices had been developed under the OSRD and were ready for service. After they were put to use in batteries on the east coast of England the percentage of buzz bombs destroyed rose from 10 to 50, then 80, and in the case of one battery, 100. I remember Dad saying that unlike the piloted bombers that went after specific targets, these unmanned missiles were addressed "To Whom It May Concern."

My father enjoyed his work during World War II. He enjoyed working with such dedicated and talented individuals, especially the scientists and military people in England. He happened to be in London on the night of April 16, 1941, which was the worst night of bombing that city ever experienced. He was moved by the courage and cheerfulness of ordinary English people under fire, and in the extensive diary he sent back to the home office in New York he commented on their "carrying great burdens with modest gallantry."

On my father's desk in his study throughout the war there stood two impressive looking shells which I have always assumed were examples of the ones fired by the anti-aircraft guns his team was working to improve.

I still have them. They're two inches in diameter at the base and over a foot tall, and on a copper band near the business end is etched, among other more mysterious numbers, "37 MM-M54." Google to the rescue! According to Wikipedia, these are indeed 37-millimeter shells like those used in the automatic canons, including the M54. A former cleaning lady of mine called them "the bullets," and refused to touch them.

I also have a whole bag full of medals Dad won over the years. After the war his contribution was recognized by three governments: He was awarded the United States Medal for Merit, the King's Medal for Service in the Cause of Freedom from Great Britain, and was made an *officier* in the French Legion of Honor, the highest honor France can confer on a foreigner.[6]

After the war he was able to devote his full attention to the Rockefeller Foundation's agricultural program, which was already well under way. The program started with efforts to improve the quality of corn and beans in Mexico and eventually expanded to Latin America and later to Asia, where the primary emphasis shifted to rice. As usual, my father traveled widely in connection with this work, and he took great satisfaction in its success.

Before the war he had been aware of the early research in the field of computers, and during the war, as we have seen, he was constantly dealing with computer design problems. He had always been interested in communication and language, and perhaps because of the unusual amount of time he spent traveling all over the world, he became increasingly interested in foreign languages and especially in the problem of translation.

Sometime in the mid-forties Dad began pondering the possibility of using high-speed electronic computers to translate from one language to another. He felt that such a procedure would be of great service to the world even if it did not produce elegant prose.

In 1947 he wrote a letter to the mathematician and linguist Norbert Wiener, the founder of cybernetics, the science of communication and control theory, proposing his idea. Wiener was skeptical. Nevertheless, Dad continued to mull it over, and in July of 1949, while on vacation, he wrote a thirteen-page memorandum expanding on his original idea.[7]

He began by recalling a war anecdote about a couple of mathematicians who shared an interest in cryptography. Mathematician A asked Mathematician B, who was bilingual in English and Turkish, to cook up a coded message on which he could try out a new code-breaking system. So Mathematician B wrote out a message in Turkish, reduced it to a column of five-digit numbers, and gave it to Mathematician A to decode.

Mathematician A came back the next day scratching his head and confessing that he had been unable to break the code. But the sequence of letters he had come up with, when correctly broken up into words, turned out to be the original message. He had decoded it properly without knowing Turkish and without knowing that the message was in Turkish.

Next, Dad recalled a well-known incident in World War I when it took allied cryptographers months to discover that a captured message was coded from Japanese, but they had no trouble deciphering it once they knew what the language was.

These stories suggested to my father that the decoding process must make use of certain frequencies of letters and letter patterns which were to some extent independent of the language used. This in turn led him to suspect that "there are certain invariant properties which are . . . to some statistically useful degree, common to all languages."

He went on to suggest four different ways of attacking the translation problem. He sent the memorandum to "twenty or thirty" linguists, logicians, and mathematicians, in the hope that it might serve as a stimulus to "someone with the techniques, the knowledge, and the imagination to do something about it." In his covering letter he wrote, "I have worried a good deal about the probable naïveté of the ideas here presented; but the subject seems to me so important that I am willing to expose my ignorance, hoping that it will be slightly shielded by my intentions."

In his autobiography he writes that the first reaction to his memorandum

. . . was almost universally negative. A distinguished linguistic scholar later told me that after the initial reading he threw the paper in the waste basket; but waking in the night and thinking about it, he rescued the memo from the basket. Over the subsequent ten or fifteen years that scholar devoted a good part of his energy to the problem of machine translation.

A few years ago a friend sent me an article about the history of machine translation that was published in 2000 in the computer magazine *Wired*.[8] The author of the article, Steve Silberman, pays homage to "Warren Weaver's memorandum" and calls Dad "the father of machine translation" (now known as MT). He also calls my father "a brilliant conceptual mathematician" (a description with which Dad would certainly have disagreed) and credits him with coining the term "molecular biology" (something for which he might have agreed to take partial credit).[9]

But Silberman is surely right in zeroing in on the most inspired idea in the memorandum, and one so characteristic of my father: his intuitive sense that there may exist certain basic elements that are common to all languages, and that this "universal interface between languages," as Silberman calls it, could be used to tackle the problem of translation. Dad wrote:

Think, by analogy, of individuals living in a series of tall closed towers, all erected over a common foundation. When they try to communicate with one another they shout back and forth, each from his own closed tower. It is difficult to make the sound penetrate even the nearest towers, and communication proceeds very poorly indeed. But when an individual goes down his tower, he finds himself in a great open basement, common to all the towers. Here he establishes easy and useful communication with the persons who have also descended from their towers.

Thus may it be true that the way to translate from Chinese to Arabic, or from Russian to Portuguese, is not to attempt the direct route, shouting from tower to tower. Perhaps the way is to descend, from each language, down to the common base of human communication – the real but as yet undiscovered universal language – and then re-emerge by whatever particular route is convenient.

Silberman writes:

Weaver's memo acted like a seed crystal dropped into a solution supersaturated with nascent ideas about computing, communication theory, and linguistics. Within two years, MT programs had been launched at MIT, UCLA, the National Bureau of Standards, the University of Washington, and the Rand Corporation. . . . It [MT] sparked an explosion of interest in formal linguistics, just as Noam Chomsky was publishing his revolutionary theories that certain fundamental structures of language were inborn.

Although there was a twenty-year period when MT fell out of favor, mainly because the goal of FAHQT ("fully automatic high-quality translation") proved to be unrealistic, by the late nineteen eighties it resurfaced in Japan and Germany.

An early speech-to-speech system called Janus that translates from English or German to English, German, and Japanese uses an interlingua approach reminiscent of Dad's metaphor of descending to the ground floor of the Towers of Babel. As he anticipated, MT turned out to be especially useful in translating technical material where the vocabulary is limited, the syntax is simple, and the problem of the multiple meanings of words does not generally arise.

In the sixties, when I began translating books from the French, I soon discovered – somewhat to my surprise – that technical language, where the words have precise, agreed-upon meanings, as in sociology or astronomy, was easier, if duller, to render into English.

My father had always been fascinated by words, language, and communication. He carried a little black notebook in which he recorded interesting oddities of language: words with opposite meanings ("fast" can mean both "rapid" and "motionless"); words that are pronounced differently when capitalized ("polish" and "Polish"); words that have opposite meanings when one letter is omitted ("public" and "pubic"), and so forth.

Information theory was a special interest, one that he shared with mathematician Claude Shannon of the Bell Telephone Labs, who had worked on fire control and cryptography for the NDRC during the war.

One day in 1949 Chester Barnard, then president of the Rockefeller Foundation, asked my father whether he had read an article by Shannon on the mathematical theory of communication in the *Bell System Technical Journal*. Dad said that he had. Barnard wanted to know whether he understood it. Dad said that he did. So Barnard asked him if he could explain the theory in "less mathematical terms," and my father agreed.

Making complex ideas understandable to the general reader was one of Dad's fortés. His "translation" of Shannon's article led to the publication of *The Mathematical Theory of Communication* (1949, University of Illinois Press) by Claude Shannon and Warren Weaver, a classic text of information theory which has been continuously in print for over sixty years.

I knew nothing at all about this until many years later; but when I was a graduate student at Oberlin, I met a geeky guy who was positively overcome when he discovered whose daughter I was. "Your father is WEAVER?"

Helen circa 1944

Chapter Four
A WOMAN OF LETTERS

There is no frigate like a book
To take us lands away,
Nor any coursers like a page
Of prancing poetry.

Emily Dickinson

The Christmas I was thirteen, Mother gave me my first diary. It was one of those chunky little "A Page A Day" affairs provided with a lock and a key, and it was such a success that it became a Christmas tradition.

I still have three of these bulging bundles of teenage angst. One page was often not enough to record the many – trials and occasional triumphs – of those far-off days. My high school diaries for 1945, 1946, and 1947 are stuffed with extra pages to accommodate the spillover of a particularly intense event (often a movie), plus lists of crushes, pictures of movie stars, poems, notes passed in study hall, doodles, drawings, words to popular songs, dreams, scraps of paper that "he" had touched, "his" entire schedule of classes, Merle Oberon's long angry speech from *A Song To Remember*, my friend Eleanor's favorite passage from *Gulliver's Travels* – even a few attempts at fiction.

My first poems echoed the accents of my beloved Emily Dickinson without her tough irony:

For every joy unbounded
A sorrow I must face;
And for each tiny triumph pay
A shame or a disgrace.

For every silver glimpse of truth
A corridor of doubt;
To balance every pinnacle

With agony a bout.

Mine not the insurrection;
Mine not the query "why?"
But merely the fulfillment of
The ratio x to y.

– Or –

Why is it when I'm all alone
The brilliant answers come

But when I see him face to face
My brain goes blank and numb?
And after every interview
The things I should have said
Come trooping in to taunt me
Now that the chance has fled?

And so on.

  By the ninth grade I had more or less abandoned art, for a very base reason. When I graduated from Greenacres Elementary to Scarsdale High there were suddenly all these kids from other grammar schools who were better artists than me. I was no longer the best artist in my class, and I didn't like that. So I channeled my love of color and form into clothes, and my need to create into writing, at which I found I could easily excel.

  For a while I didn't show my poems to anyone except my best friends, Peggie, Eleanor, and Ingrid, and occasionally, my parents. Dad sent "Ratio" and another one called "Forever" to a poet friend of his, who was very kind.

  One day my favorite English teacher, Mrs. Bartlett, gave us an interesting assignment: to write a parody of an existing poem. My brother, who was living at home while taking a journalism course at Columbia, helped me find a likely poem, Robert Frost's "Revelation." He even gave me an idea of how to proceed.

  Mrs. Bartlett liked my parody, "Cosmetics," so much that she sent it to Frost, who was still very much alive. He probably hated it, but Mrs. B. published it in the school literary magazine, *Jabberwock*. (Our yearbook was the *Bandersnatch*: in Scarsdale, there was no escaping Wonderland.) It was my first published writing.

Encouraged by this, I started taking some of my older poems out of mothballs, and they were accepted, too. I found I had a talent for comic verse, and gained a reputation as a wit.

Funny, Dad never told me that *writers* (unlike artists) didn't make any money. Maybe because he was a writer himself; or maybe because I never told him I wanted to be a writer. Maybe I didn't know it yet.

Actually, everyone in the family was a writer. My father, in addition to at least seven books, wrote hundreds of articles on everything from "The Sedimentation of Small Particles in a Fluid" back in 1928 to his much reprinted article "Can A Scientist Believe in God?"

My brother, Warren Weaver Jr., eventually became a highly respected political reporter at the Washington bureau of the *New York Times* who turned out textbook *Times* copy under his own byline.

And my mother, who read widely all of her life, was a great letter writer. She liked to write first thing in the morning when her mind was fresh. She wrote with clarity and precision but also with a warmth and intimacy that did not come easily to her otherwise. Mother always said you could say things in letters that you couldn't say any other way.

My love of words, like my father's, extended to foreign languages, and in the ninth grade I embarked on four years of Latin and four years of French.

In choosing those courses over other electives like chemistry or physics, I was picking subjects I knew would be easy for me, and avoiding the more challenging ones that might lower my grade point average. High marks were important to my father, who rewarded me at the end of every marking period by giving me ten cents for every point over ninety: a 98 in English earned me eighty cents, and so forth. He joked that he was going to *deduct* ten cents for every point I got over 75 in gym: lack of athletic ability was one of our great bonds.

At first I thought getting money for high marks was silly, but after a while I started to like it. Since I was not a social success – I didn't date or go to dances, couldn't throw a ball or do the Lindy – being a "brain" was about the only identity I had. I figured I might as well accept it.

In any case, I took to the rigors of Latin with Mother's help, and as for French, it came to me so easily that it was almost as if I were remembering it. I've often wondered if my lifelong love affair with that language and the ease with which I learned it had anything to do with hearing it all around me in Paris as an infant. The summer I learned to walk in Saint-Cloud I was surrounded by people who cooed at me in French and called me "*chère petite choux*" ("dear little cabbage," a common term of endearment). The accents of that tongue were familiar to me and were associated with happiness, innocence, and success. Then there was the great good luck that two out of my three best friends spoke French.

Peggie's father, George Bakeman, was head of the Paris office of the Rockefeller Foundation. The Weavers met the Bakemans that first summer in Paris. George was from Newton Upper Falls, Massachusetts, but his wife Mollie was an eccentric Russian émigré, and their three daughters were born in Vienna, London, and Paris. The two families hit it off from the start, and since Mother and Mollie were both great letter writers, stayed in touch across the miles.

The Bakemans got out of Paris just before the German occupation. After a brief stay in Flushing, New York, they bought a farm house in Richmond, Virginia, a former Italian gambling den which Mollie, with intentional irony, dubbed Malmaison, after the palace of the Empress Josephine near Versailles.

By this time, due largely to the flourishing correspondence of the two mamas, our two families had become great friends. The Bakemans visited us in Scarsdale, and during summer vacations, Peggie and I exchanged visits between Connecticut and Virginia.

Although two more different women could hardly be imagined than my modest and reserved mother and the irrepressible and self-dramatizing Mollie, they became best friends for life. And against all the odds Peggie and I, who at best were expected to play nicely together while our parents lingered at the dinner table roaring over my father's and Mollie's stories, also became best friends for life.

Everyone in the family except George, "the New Englander," had some kind of exotic foreign accent. Peggie was a bright-eyed, pigtailed, enthusiastic playmate who was ready for anything. In the eyes of this little refugee who was, as her mother put it admiringly "*contente de si peu*" ("happy with so little" – i.e., not spoiled, like me), I was the poor little rich girl – poor, because I was pale and sickly, and didn't appreciate what I had. Growing up in Scarsdale where I was odd man out, this breath of fresh air from Europe helped to jolt me out of my romantic attachment to victimhood and provided another point of view.

Peggie and I shared a love of poetry, especially Dickinson, Whitman, and Poe. We copied their poems into our journals, and on one of my visits to Richmond, we made a pilgrimage to the Poe Museum. By the time we were teenagers, we were carrying on a massive correspondence of our own, partly in French.

In the middle of my sophomore year at Scarsdale High another breath of fresh air from Europe blew into my life. Born in Manilla to a Swedish father and a Yugoslavian mother, Ingrid had lived through the Japanese occupation. Shortly after her arrival she startled everyone in my biology class by diving under her desk at the sound of an airplane roaring overhead. Ingrid was annoyed to have made a spectacle of herself; but I loved her for opening my eyes to a world beyond Scarsdale. I knew at once that I wanted to be her friend.

Ingrid was big, blonde, and beautiful, with slanty green eyes and a smile you couldn't resist. Everything about her was different, including her clothes. In lieu of the regulation sloppy joe sweaters, bobby socks, and loafers with pennies, Ingrid wore whatever she had grabbed on her travels, which included peasant blouses and some very unsloppy short-sleeved sweaters. She wore a man's watch on her wrist and her hair was often up in braids. Her unconventional appearance was a source of eyebrow-raised wonder to the socially elite of the school which, needless to say, did not include me.

Ingrid and I discovered a mutual love of music, poetry, and the French language and our friendship blossomed overnight. We wrote poetry together and invented a kind of surrealist art form called Glumphs. Best of all, she couldn't care less what anyone thought of her. As a teenager in Scarsdale, that was unheard of.

Her French was so good that Mr. Reid accepted her into the honors class in the middle of a marking period. We spoke French together and became more and more at home in that language. It was very useful for passing notes in Study Hall about boys we had crushes on when maintaining secrecy was essential.

In my diary I wrote, "Iggie's just as nuts as I am, and it's such a comfort." I would probably have loved French, no matter who taught it, but there couldn't have been a better teacher of French – or of anything else, for that matter – than Charles L. Reid, Jr. A slight, sallow-complexioned little man with a greying crew cut, an undernourished moustache, and impeccable clothes, Reid had a reputation for being the deadpan wit of the faculty. His dry humor and clear explanations made even French grammar absorbing, and his focus on pronunciation was most unusual for a high school language instructor. He was not above touring the classroom while we practiced our r's and nasals and feeling the jaw muscles of every student in the class to make sure we were expending the required amount of energy. For Americans, he insisted, were lazy; speaking proper French demanded effort: *"Il faut morder les syllables!"* ("You have to bite off your syllables!")

His relentless drilling on pronunciation produced amazing results with the most blatant of New York accents, and though he was serious about its importance, he could always make us laugh. When he didn't like something, it was, *"Oh – ça me shocke!"* He could be sarcastic on occasion, but his seemingly sinister exterior was really an act. Amid the chaos that reigned in Room 310 before he appeared, we would hear his voice preceding him down the hall and giving us fair warning: *"J'arrive!"*

He insisted on giving us a sense of French history, literature and culture. From him I first heard about the Dreyfus Case, which I vaguely remembered from having seen a 1938 movie called *The Life of Emil Zola*.

Reid taught us how to read poetry, something no previous teacher of mine had ever attempted. In those days students were required to memorize poems, something I hated at the time (because we had to recite them in front of the class) but appreciate in retrospect, and am sorry to see no longer done.

He introduced us to the other arts, too, bringing in reproductions of Impressionist paintings as well as records of Franck, Ravel, and Saint-Saens. He took us to the Cloisters to see the unicorn tapestries and into Manhattan for French movies, when they were just becoming fashionable.

Mr. Reid pretended to despise all extracurricular activities except his own French club, but this may have been another of his poses. He did genuinely despise athletics and refused to attend sports assemblies any longer than the time required to conduct those students assigned to his home room. Once they were settled, he would wait a maximum of ten minutes and then he would rise and, with an air of infinite disdain, would glide up the aisle and out of the auditorium.

In spite of his attempts to stay in the character of intellectual snob, Mr. Reid could not conceal his real affection for our class. It showed through in the way he called us "*mes enfants*," and the solicitude with which he primed us for distinguishing ourselves in college. He groomed me for a national competition, held my hand after I threw up in the girls' room before the oral part, and was delighted when I came away with a "*médaille d'honneur*." In later years he referred to our class of '48 as "My Class," or "M.C." for short.

When I took up translating books from French as a career and actually made a success of it, it had as much to do with my extraordinary good luck in having Charles Reid for a teacher as it did with my lifelong friendship with my other mentor, Richard Howard. When, in 1976, my translation of the Artaud *Selected Writings* was nominated for the National Book Award, I sent Mr. Reid a copy, gratefully inscribed.

For years I sang in the choir at the First Congregational Church, and senior year I signed up for Chorus at school. I had given up the piano because practicing interfered with my grades, but I still had music in my life. I got to sing Handel's *Messiah* at Christmas and *The Gondoliers*, the glorious Gilbert and Sullivan operetta, in the spring. Our director, Mr. Lawson, wanted me to sing one of the high soprano recitatives in The Mess, as we called it, and to try out for a principal part in the operetta, but I was much too nervous to even consider it. When it came time to plan our graduation ceremony, though, I did agree to sing first soprano in an octet.

Then in June of my senior year I got the mumps, and it looked like I wouldn't be able to sing after all. While I lay in bed, my father read to me. I was almost seventeen, but what I wanted to hear was Winnie the Pooh.

In the last chapter of *The House at Pooh Corner* Christopher Robin has to go off to school and learn about fractions and things, and he has to say goodbye to his bear. The last paragraph goes:

So they went off together. But wherever they go, and whatever happens to them on the way, in that enchanted place on the top of the Forest, a little boy and his Bear will always be playing.

My father got through the enchanted place on the top of the Forest, and then he simply couldn't go on. He had to put the book down without finishing. It didn't matter, because we both knew those words by heart.

I remember hating how I looked with my jaw all puffed out at the sides and my face pear-shaped, like Kate Smith. This went on for so long I started to think I was always going to look this way. One day I decided to accept my new face, and get on with my life. The next day, the swelling was gone. It seemed like some kind of a lesson. I recovered from the mumps in time to sing at my graduation. I remember how it felt to stand on that stage with Nina and the others and hear my voice soaring over that big room, right on pitch.

I graduated fifth in a class of 321 – much to my relief. I had dreaded the possibility of being valedictorian and having to make a speech.

In the yearbook under my picture it said, "No genius without a little madness." I always liked that. It made me feel that after all, and in spite of everything, I had at least been understood.

Helen and her Dad, Warren circa 1950

Chapter Five
PROBABILITY AND ESP

You do ill if you praise, but worse if you censure, what you do not understand.[10]

Leonardo da Vinci

I couldn't have been much older than seven when Dad asked me, "If something seemingly impossible happened – something that seemed to be a miracle – do you think it would mean the temporary suspension of one of the laws of nature, or the existence of a law that was previously unknown?"

I thought it over and chose the latter. He seemed satisfied with my choice.

One of my father's areas of expertise was the theory of probability. Indeed, his very first published writing was a little article on this subject entitled "Chance" that appeared in *The Wisconsin Engineer* in 1916.[11]

In it Dad explains that the laws of science depend on the predictability of cause and effect: "a sure and substantial knowledge that a given set of conditions will result in certain definite events." But there is a large class of phenomena for which the causes are unknown. In these cases we are (or we think we are) dealing with chance. Probability deals with the *laws* of chance. For example,

If one tosses a coin into the air, it is impossible to say whether it will fall heads or tails. . . [yet] while no man can say whether a single trial will offer a head or a tail, it is certain that [if we keep tossing] eventually we shall get half heads and half tails. This expresses one of the greatest laws of probabilities, Bernoulli's law of great numbers.

Of course, it is always possible that at some later date causes that have been obscure may come to light, and that new knowledge, made possible by new analytic tools, may thus elevate an event from the field of probability to that of those sciences which treat with certainty of a single case. . ... Every event is the result of the laws of nature, the causes for some being obscure merely because we are ignorant of the ties which unite such events to the entire system of the universe.

This last sentence provides an answer to the question my father would put to me when I was seven. Twenty-two years later, my father was still fascinated with the idea of seemingly impossible events and their relation to science.

In "The Reign of Probability" (1930) my father makes a distinction between unitary laws, which describe the behavior of individual units, and statistical laws, which describe the *probable* behavior of large groups of units. Statistical laws are probability laws.

But after reviewing Eddington's theory of time, recent advances in quantum physics, and Heisenberg's uncertainty principle, Dad concluded that the difference between unitary laws and statistical laws is only apparent. Everything, he decides, is subject to probability:

It seems impossible to escape from this mathematical goddess of chance. . . such is our relationship with the external world that all our data are of necessity probability data.[12]

In 1948, in an article entitled "Probability, Rarity, Interest, and Surprise," he explains the difference between all these concepts and concludes that if you ever happen to be present at the bridge table when a hand of thirteen spades is dealt, remember that what you ought to say is this: "My friends, this is an improbable and a rare event: but it is *not* a surprising event. It is, however, an interesting event."[13]

"Not surprising" to an expert in probability, perhaps, since all hands are equally likely; but try telling that to the player who picks up the hand!

Even when highly improbable things happen – say, a run of seventy-five heads in a row when tossing a coin, or a hand of thirteen spades – this does not disprove the theory of probability:

Weaver: TRANSLATION OF LIGHT

For the event that has now occurred, although amazingly rare, is still an event whose probability can be calculated, and while its probability is exceedingly small, it is not zero. . .. And even if miracles persist in occurring, these would be, from the point of view of probability, not impossible miracles.

Thus in a strict sense probability cannot be proved either right or wrong. But this is, as a matter of fact, a purely illusory difficulty. Although probability cannot be strictly proved either right or wrong, it can be proved useful. The facts of experience show that it works.[14]

Dad's little book *Lady Luck: The Theory of Probability* (Doubleday, 1963 and Dover, 1982) is often described as a popularization of probability theory. The first few chapters, where he traces its disreputable origins in gambling houses and other dens of iniquity, are written in a charming and readable style. But all too soon the mathematical formulas take over; and although I was always pretty good in math (and of course had the finest possible help with my math homework), Dad lost me on page 87 with permutations, factorials, and combinatorial analysis.

I should have hung on longer, because the book is not all formulas. For example, one of Dad's favorite probability stories was the famous (among mathematicians, anyway) legend of the monkeys at the typewriters. In *Lady Luck* he quotes this extract from an address given before the British Association for the Advancement of Science:

If six monkeys were set before six typewriters it would be a long time before they produced by mere chance all the written books in the British Museum; but it would not be an infinitely long time.[15]

Dad comments that that august British body "is sometimes referred to, in England, as the British Ass."

After *Lady Luck* was published, my father wrote several mathematician friends in an effort to track down the original source of the monkeys idea. (In some versions, all the monkeys had to do was sooner or later, type out *Hamlet*.) Dad eventually traced it back to a 1914 book entitled *le Hasard* by French mathematician Émil Borel.

Monsieur Borel discusses "*le miracle des singes dactylographes*" (the miracle of the typing monkeys) and comments that however improbable, the outcome is "logically possible," and that this "myth of the typing monkeys is not without profound philosophical significance."

Lady Luck has been translated into nine languages: British, Danish, German, Hebrew, Polish, Romanian, Spanish, Swedish, and Japanese, and fifty years later, the American edition is still in print. But the nicest tribute to it that I know of is in a letter Isaac Asimov wrote to their mutual publisher:

As for the book by Weaver, listen – it is the sanest discussion on probability I have ever seen. Explaining Gaussian distribution so that the non-mathematician can understand it is sheer genius.[16]

A certain open-mindedness on my father's part about subjects that seem to fly in the face of science is illustrated by his attitude toward dowsing. In an article entitled "Mathematics and the Problem of Ore Location" (1930) he quotes a sixteenth-century treatise on mining by Georgius Agricola that described the use of divining rods in the location of ore. Agricola considers the evidence for and against the forked twig and concludes, "Since this matter remains in dispute and causes much dissension amongst miners, I consider it ought to be examined on its own merits." My father comments, "His truly scientific spirit, and the attitude proper for us, are alike indicated by this ancient author's remark."

This same open-mindedness can be seen in his personal and professional relationship with psychologist Joseph Banks Rhine (1895-1980), the pioneering researcher in the field of extrasensory perception. At least as far back as 1936 the two men were in communication, and as an officer of the Rockefeller Foundation my father's support was decisive in getting funding for the work of Rhine's Parapsychology Laboratory at Duke University in Durham, North Carolina.

ESP was one of my father's obsessions. Although he did not believe in it, he did not rule it out, either, and he was impressed when Rhine and his associates began piling up statistical evidence in its favor. Dad said the whole subject made him so nervous he didn't like to think about it, but he knew that wasn't the right attitude for a scientist.

In *Lady Luck* he comments,

Weaver: TRANSLATION OF LIGHT

A group at Duke University, under the leadership of J. B. Rhine, is convinced that the mind is capable of exerting a direct influence on physical events. . . . Although it is difficult to refute their evidence, this result has not generally been accepted by scientists.

. . . In the book cited [J. B. Rhine and J. G. Pratt, *Parapsychology*] there are described experiments which are interpreted to prove that a person can be "aware" of what is in another mind (telepathy); can "perceive" objects or events at a distance (clairvoyance) without any use of the senses of sight, hearing, touch, etc.; and indeed that the mind can directly, and without any other than a purely mental effort, have a material effect on a physical system (psychokinesis – for example, "willing" that one face of a die come up significantly more than one-sixth of the time).

Commenting on the conclusions Rhine draws from certain experiments (which resulted in a 0.0005 possibility that the results were due to chance), Dad writes,

> On the one hand, we are asked to accept an interpretation that destroys the most fundamental ideas and principles on which modern science has been based; we are asked to give up the irreversibility of time, to accept an effect that shows no decay with distance and hence involves "communication" without energy being involved; asked to believe in an "effect" that depends on no known quantities and for which no explanation has been offered, to credit phenomena which are subject to decline or disappearance for unexplained and unexplainable reasons. On the other hand, we are asked not to believe that a highly improbable chance result has occurred. All I can say is, I find this a very tough pair of alternatives.
>
> The Rhine ESP results could be explained on the grounds of selection or falsification of data. Having complete confidence in the scientific competence and personal integrity of Professor Rhine, I find this explanation unacceptable to me.
>
> As I have said elsewhere, I find this a subject that is so intellectually uncomfortable as to be almost painful. I end by concluding that I cannot explain away Professor Rhine's evidence, and that I also cannot accept his interpretation.

Dad's open-mindedness was part and parcel of his understanding of probability. The line between chance and causation is very difficult to draw.

In a letter to *The Scientific Monthly* in February 1950 he wrote:

> . . . as long as twelve years ago, Professor Rhine and I discussed together how it might be possible to get some persons to stop wasting time attacking certain actually unassailable statistical aspects of his work, and how to get the scientific community to direct its skill and energy to those phases of the problem that deserved and required attention.[17]

In August of 1950, the Rockefeller Foundation made another grant to the group at Duke.

In 1959 Dad wrote a mathematician friend at Princeton a letter that started out, "Every once in a while I wake up in the night and worry about ESP!"

And a year later, in a letter to his old friend and mentor Max Mason, he said he was going to Duke to spend "the better part of a day" with Professor Rhine. "I haven't seen him or visited his laboratory for a long time. And I do not seem to be able to succeed in forgetting about ESP."

After that visit, he wrote Rhine a seven-page letter analyzing the evidence for his findings in terms of probability and statistics. He starts out:

I gather that you and your immediate colleagues consider that the existence of four types of psi [psychic] phenomena (*telepathy, clairvoyance, PK* [psychokinesis], and *precognition*) has been experimentally established. But I think you also recognize that very many other persons have not accepted this fact. Some actively reject the conclusion and others (many more, I suspect) prefer to disregard the evidence rather than to weigh it and either accept or discard it.

And after detailed analysis of his findings – refuting some points and accepting others, and advising Rhine how his language could be made more precise and thus his conclusions more acceptable – he asks for a confidential statement of his budget and wants to know what use he would make of an additional grant from the Rockefeller Foundation. (My father had retired from the Foundation the year before, but he was still concerned that they continue to fund Rhine's work.)

Whenever he came across an example of unexpected synchronicity – say, letters crossing in the mail – Dad would say, "By George, Professor Rhine must have something!"

After my father died I found three of Rhine's books in his professional library, one gratefully inscribed, and another containing a final letter dated March 15, 1973 thanking my father for his help over the years.

Dad loved order, reason, and logic; but he was also drawn to their opposites. For example, he was intrigued by the world of poetry and dreams. Come to think of it, what is *Alice in Wonderland* in the end but one long and unforgettable dream? The story, the characters, the language, and the whole atmosphere of my father's favorite book are but the figments of a dream. Its logic is the logic of dreams.

Dad collected examples of mathematical and scientific breakthroughs that came about through dreams. I've mentioned two so far, and I remember another that particularly impressed us both.

Sometime in the sixties I was lucky enough to be invited to join my parents at dinner with Jonas Salk and his wife Françoise Gilot (who had also lived with Picasso). Somehow the subject of dreams came up.

Salk, the great microbiologist who is generally credited with discovering the polio vaccine, said that he kept a tape recorder next to his bed, and that when he woke up in the morning, he recorded his dreams. He said that he got ideas, information, and solutions that helped him with his work in his dreams.

Helen in Greenwich Village circa 1956

Chapter Six
REBELLION[18]

The only people for me are the mad ones, the ones who are mad to live, mad to talk, mad to be saved, desirous of everything at the same time, the ones who never yawn or say a commonplace thing, but burn, burn, burn, like fabulous yellow roman candles exploding like spiders across the stars and in the middle you see the blue centerlight pop and everybody goes "Awww!

Jack Kerouac, *On the Road*

When it came time to go to college I applied to Swarthmore in Pennsylvania, Middlebury in Vermont, and Oberlin in Ohio and was accepted at all three schools. When I decided on Oberlin it was not only because it had a famous conservatory and I knew I'd be surrounded by good music. It was also – mainly, I think – because I wanted to get as far away from Scarsdale as possible: Scarsdale with its cliques, its Golf Club, its Republican majority, its "what does your father do?" I doubt if I was even aware of Oberlin's long and distinguished history of political liberalism. But I did find a lot of other oddballs, brains, and misfits there, and felt right at home.

In getting away from Scarsdale I was getting away from my parents. My struggle for independence turned out to be a struggle with Dad. It's hard for me to remember why I was so angry at my father in those days. All I can come up with is that I felt as if, even at a distance – or maybe especially now that I was no longer under his roof – he was trying to control me.

As long as I can remember, I've been drawn to solitude. I have liked my own company. Whether it was talking to my stuffed animals or cutting out paper dolls or listening to the radio or reading or writing in my diary or just daydreaming, I always had things I wanted to do by myself. I had good friends, but I needed an enormous amount of time alone to digest my experience. I had the writer's temperament long before I became a writer.

In high school there was always a vast amount of reading that had to be done and not enough time in which to do it. Partly because of my poor health, my parents enforced a strict curfew which I regularly violated in order to keep up. I'd read in bed with the book under the covers and one hand on the light switch in case Daddy came upstairs to check on me. And I always kept my door closed when I was upstairs in my room.

On a typical evening, my parents would be down in the living room listening to the radio, reading aloud, or talking. At some point my father would come upstairs, open my door a crack, and ask me what I was doing. I would answer him, probably with an ill grace, and he'd go back downstairs – but he'd leave the door to my room ajar. I'd wait until he was back in the living room and then I'd get up and shut the door. He wanted some sort of connection, no matter how small. I wanted to be left alone.

This dialogue of the door went on for years.

It was not until I began living in the dorm that it began to dawn on me how confined and protected I had been in comparison with my classmates.

Most of the girls in Tank Hall were given a sum of money for their expenses for the term, and if they spent it all in the first few weeks, too bad for them. I was sent a check for $32.50 every month, and expected to account for every penny of it. When it was determined that I couldn't be trusted to do this, my father started sending me $16.25 every two weeks. I was expected to write home every Sunday, and if I failed to do this, my parents became seriously alarmed.

I was the only girl in my dorm whose mother did her laundry. There existed in my family, probably a relic from the time my brother had attended summer camp, a large box made of some sturdy pre-plastic material with a tight-fitting lid secured by canvas straps and a little metal-bound window in which you inserted a reversible address card. Instead of taking my sheets and towels down to the ancient washing machine in the bug-infested basement of Tank Hall or to the local laundromat, like everyone else, I put them in this box and sent them home to Mother in New York. A week later, I'd get the box back with everything neatly washed, ironed, and folded, with fragrant apples and oranges, crackers and cheese tucked into the corners.

When I think of that box now, arriving at the dorm so full of my mother's orderliness and love, I am tempted to believe that my parents did not prepare me for life in the real world. In any case, the contrast between the sheltered life I had lived at home and the freedom of life at school was dramatic and disorienting. I began to resent my parents for "overprotecting" me.

I began questioning their values. I had never been particularly political, but in Archibald Byrne's freshman English Composition course I was exposed to liberal ideas. I began spouting them in a half-baked manner when I came home for the holidays, much to the disgust of my father.

Suddenly I had this reservoir of resentment against my father that felt bottomless.

I remember sitting at my desk in my room in Scarsdale during spring break and my father coming to the door to remind me that it was time to help my mother set the table. For some reason, this infuriated me and I began screaming at him, something unheard of in our family. My screams escalated into a full-scale tantrum. The next thing I knew I was lying at the top of the stairs pounding my fists on the floor and sobbing. This whole performance – for it was at least partly that – scared my father so much that he backed off without the usual lecture.

Things got so bad that whenever it was time to go home from college, I promised myself that I would try to get along with my father. The truce would last about twenty-four hours, and then we'd be at loggerheads again.

When I brought home my first actual boyfriend the summer after my freshman year, Dudley charmed my mother, but my father was suspicious. He told me that Dudley and I could not speak to each other when we were on the second floor of the house – presumably, because that's where the bedrooms were.

Dudley had the highest record for cutting classes in the history of Oberlin College. He spent a lot of time in the pool hall, sharpening his skills and drinking 3.2 beer, which was all you could get in Ohio in the fifties. Mother said that being a good pool player was the sign of a misspent youth.

He was my first rebel.

I had been introduced to my parents' Victorian attitudes toward sex at an early age. When, at about age five, I asked my mother where I came from she said nothing, but what she did spoke volumes. She was sitting at her dressing table. She put down her hairbrush and placed her right index finger on her lips in the universal command for silence. Meanwhile her left index finger made jabbing motions toward her crotch.

That was it! That was Mother's Sex Lecture. Except for some practical facts about menstruation, that was all the information I got from my family about sex.

I first heard the boys saying the F word one day in the fifth grade, but nobody would tell me what it meant. As soon as I got home from school I went to my father's study, pulled the big dictionary onto my lap, and turned to the F's. But in Webster's Unabridged, Second Edition, the final authority in hair-splitting arguments at the dinner table, "fuck" was nowhere to be found.

At the dinner table that night my parents, my brother, and I sat sipping Campbell's Cream of Tomato Soup. I waited for a gap in my father's monologue. When it came I turned to him and said in as casual a voice as I could muster, "Daddy, what does 'fuck' mean?"

Suddenly, there was soup everywhere. My father was choking, sputtering, dabbing at his mouth with his linen napkin. When he could finally speak he put down his napkin and said sternly, "Young lady, you are not to use that word in this house!"

"But what does it *mean*, Daddy?"

"The subject is closed."

So that was that. Even though I had only the vaguest idea what the word meant, I liked the sound of it. It had an undeniable impact. According to the immutable law whereby anything forbidden automatically becomes irresistible, the word I didn't know the meaning of became my favorite word. It became my mantra, which I chanted under my breath when I needed to let off steam.

The summer I was thirteen a boy walked me home from the beach one day in broad daylight. My father told me I was too young to have a boy walk home with me. This was a perfectly harmless Irish Catholic boy who actually bored me to death. (The boy I liked was from the "wrong side of the tracks" and it was rumored that he had once been seen necking with his stepmother on the beach.)

Dad was invariably suspicious of my boyfriends. He didn't much care for the man I married, either, an art history major from Brooklyn I dated my senior year at Oberlin. I married him for the worst possible reasons: to get away from home and to lose my virginity – two items that were practically synonymous – but most of all, because for all intents and purposes, my whole life had been school, and I had no idea what I wanted to do after graduation.

I graduated in June, 1952 with a major in English literature and a minor in French. Unlike my brother, who had been elected to Phi Beta Kappa his junior year and had graduated from Amherst *summa cum laude*, I didn't make Phi Beta Kappa until senior year and only earned a *magna*. My new goal in life was to have a good time. In high school I had been a sort of disembodied brain; from now on there seemed nothing more important in life than becoming a complete human being.

My father joked that he did not really expect my husband-to-be to come to him and say, "Sir, I crave the hand of your daughter in marriage." But I could tell he would have liked some sort of acknowledgment that he was, or had been, in charge.

I was married in August in the chapel of the First Congregational Church in Scarsdale. It was a chintzy little wedding, by my choice: I wore a dress I had worn as a bridesmaid at my brother's wedding the year before; there was no engagement ring, and I bought the wedding ring myself. My father took the photographs. I've often wondered what he thought when he saw that blatant sexual leer on my face as I looked into my young husband's eyes just before we boarded the train for our honeymoon. This event took place in Washington, D.C. – my parents' idea. In our nation's capital, the Earth did not move.

My father went through the motions of giving me away but looking back, I think my parents must have sensed that this was a non-event. Three years later, when I announced that the marriage was over, they didn't seem that upset or surprised.

I have always looked upon the year of my divorce – 1955 – as the real beginning of my life. I moved to New York's Greenwich Village and more or less went wild.

My first apartment was a third-floor walkup on Sullivan Street. After I got settled I found a job as a "Gal Friday" for a small book publishing company on lower Fifth Avenue.

My father said this was the wrong way to go about it: I should find a job first, and then look for an apartment; but since he gave me the money for the first month's rent plus the month's security anyway, I did it my way. That was Dad: forever telling me that money didn't grow on trees, and then undermining his own argument by his never-failing generosity. Even after I started working he continued to send me a monthly allowance of $50. From time to time I would try to wean myself off of this nice little cushion, and my clumsy attempts to do so often hurt his feelings. He said he wanted me to be independent, but sometimes I got the feeling that he really wanted me to remain a little girl.

The truth was, he didn't trust me to take care of myself, and rightly so. I knew he was always there to bail me out. I counted on his help, but resented his lack of trust. We rarely discussed our differences face to face. In our family it was traditional not to express your feelings – except occasionally in letters. The worst letters I got from Dad were the ones that started off, "First of all, I hope you understand that I love you very much, and I only want what's best for you. But – "

In his letters Dad would express all his fears about my future, my lack of discipline, whatever it was that was keeping him awake at night. He'd get it all off his chest and down on paper, put it in an envelope, put a stamp on it, mail it, and feel all better. I'd get the letter and feel terrible until I had done the same. This went on for years. Meanwhile that reservoir of anger was shrinking, because for once in my life I was free to do as I pleased.

But given my need for freedom and his for control, is it any wonder that I was attracted to men who were the exact opposite of my father? Penniless artists, struggling writers, alcoholics, freeloaders, homosexuals – guys who couldn't have taken care of me even if they wanted to.

I guess the ultimate rejection of my father was being gay, which I thought I was for a while. Shortly after I moved to the Village I had a brief affair with a girl I had met at Oberlin, a piano major at the Conservatory. I even took her home to meet my parents, both of us in full drag. My innocent parents noticed nothing. I fantasized telling my father, "Dad, I'm a dyke!" but I didn't have the nerve, and anyway, it wasn't true. I soon realized that I was hopelessly heterosexual.

It was while I was living in the apartment on Sullivan Street that I met the two people who had the most profound influence on the future course of my life.

The first of these was the poet and translator Richard Howard, who became my lifelong friend and my mentor. Richard was personally responsible for my eventually making the transition from nine-to-five employee in a publishing house to professional literary translator.

The other person who changed my life was Helen Elliott, a girl I met at a luncheonette across from my publishing house, who shared my love of rock'n'roll and who, it turned out, knew everyone who was anyone in the bohemian and Beat subculture of that time.

Helen and I found an apartment on West 11th Street and moved in together. It was through Helen that I met Jack Kerouac, Allen Ginsberg, Lenny Bruce, and some other wild characters in the New York City of the fifties and sixties.

The day I met Jack and Allen, Jack had not yet found a publisher for his novel *On the Road*, but Allen's poem *Howl* was already an underground classic. The two men stood confident on the threshold of their fame and exuded a charisma that was irresistible. Jack needed a place to stay in the city. He and I fell in love on sight, and he moved in that night.

I loved Jack because he was beautiful and talented and also because he was the opposite of everything I had been surrounded by growing up in Scarsdale. He was French Canadian, and had grown up in poor neighborhoods in the rough mill town of Lowell, Massachusetts. He had those exotic New England vowels, and could sing like Frank Sinatra. Best of all, he was a writer – which was what I wanted to be myself.

Our affair lasted only a few chaotic months, but I did manage to take him home to Scarsdale to meet my parents.

I met Jack at Grand Central Station after I got out of work. He was carrying a bottle of Jack Daniels in a brown paper bag. He said it was a present for my father, but I had the distinct impression that Jack had opened it and started in on it himself.

He was nervous about meeting my father: "What's he gonna think of me?"

"They'll love you!" I assured him. But I was nervous myself.

I had picked a man who was as unlike my father as possible: a penniless writer, twice married, and a deadbeat father who was probably a hopeless alcoholic.

What, indeed, was my father going to think of him?

Against all the odds, the evening was a success.

No sooner had we walked in the door – no sooner had I introduced Jack to my parents – than he asked them whether they believed in God. They may have been startled, but they were also impressed by his seriousness, and this was a subject close to both of their hearts.

My father was delighted with Jack's gift. If the bottle had been opened he pretended not to notice, and went off to the kitchen to make Old Fashioneds for Jack and himself, Mother and I abstaining.

Soon Jack and my mother were engrossed in conversation about their religious beliefs while Dad and he proceeded to pay their respects to the Jack Daniels. Warmed by the liquor, my father temporarily forgot his distaste for the bohemian life and his anxiety about his daughter's being mixed up with this unusual person.

Jack and he had both grown up in small towns in America with fathers who had trouble making a living. They were both shy, sensitive and ambitious men who loved the books of Joseph Conrad and the sea and music – and me. Somehow, the vice president of the Rockefeller Foundation and the self-styled Dharma bum managed to bypass their differences and focus on the things they had in common. Jack forgot his nervousness and my parents were completely disarmed. I had been right: they loved him.

On the train back to Grand Central that night Jack told me how much he had liked my parents, especially my father. He had picked up on how much my father loved me. Jack, who had absolutely no use for Freud or psychoanalysis, had a flash of insight:

"You should fall on your knees and worship him! He's your *giant secret lover*!"

Analysis was the religion of the fifties, and after I moved to the city I had made a few attempts at finding an analyst. My marriage had failed and every other relationship had ended in either boredom or heartbreak. I knew I was "neurotic," but I hadn't found anyone I trusted to help me.

Around the time I moved in with Helen, I had changed jobs. I was now assistant to the managing editor at Farrar, Straus and Cudahy. I loved my new job and wanted to keep it; but living with Jack and Helen, neither of whom had to get up and go to work in the morning, was turning out to be a very stressful experience. The late hours, the lack of sleep, Jack's drinking, and his general unpredictability, were getting on my nerves. I loved him, but he was making me crazy. I knew I needed help.

There was no way I could afford to pay for psychoanalysis on my secretary's salary. The day before Jack signed the contract with Viking Press for the publication of *On the Road*, I had a drink with my father at the Algonquin Hotel. I had screwed up my courage and made a date to meet him after work. I was planning to ask him to help pay for psychotherapy with a man I had found on Park Avenue.

I was nervous. I truly believed that my "neurosis," if that's what it was, was largely the fault of my repressive upbringing, and it felt odd to be asking my father to pay to undo the damage he and Mother had done. The current thinking was that analysis was more likely to be successful if you paid for it yourself, but I didn't feel I had a choice.

So I swallowed, took a deep breath, and told him, "I can't have a healthy relationship with a man. I need professional help."

To my amazement he agreed to help me. He wanted, of course, to know this man's credentials – where he had studied, what degrees he had, and so forth. The truth, which I should have known by then, was that it was almost impossible for my father to refuse me anything I really wanted. But this was a breakthrough for us, because although the word "sex" was never mentioned, on some level my father was acknowledging my sexuality. I felt as if I'd grown an inch.

Years later I learned that the Rockefeller Foundation had underwritten Alfred Kinsey's research into the sexuality of the human male. The Foundation had continued to support Kinsey throughout the furor created by his book *Sexual Behavior in the Human Male* (1948) but had cut off his funding when he turned his attention to the supposedly virtuous American female. Kinsey published those results in 1953 in *Sexual Behavior in the Human Female.* As director of the division of natural sciences at the Foundation from 1932 to 1959 my father's was the decisive vote, but in his memorandum recommending that the Foundation cease funding Kinsey's work he made it clear that his decision was based not on "reluctance to deal with such a subject," but on Kinsey's inadequate sampling procedures and elementary errors in statistics.

But Dad did agree to underwrite my research into my own sexuality, and considering his upbringing and his attitude toward sex, I think he was amazingly generous.

A few weeks after I started analysis, I asked Jack to leave. He was bitter and blamed our breakup on my analyst, but it was my decision. (That guy never opened his mouth, except to say, "Why do you ask?" Once I even caught him snoring.)

I soon realized I needed my own space. Through a friend of Jack's, I found an apartment on West 13th Street. Jack forgave me, and we remained friends.

On the Road was published in the fall of 1957 to mixed reviews. In the daily *New York Times* Gilbert Millstein hailed its publication as "an historic occasion" and compared it to Hemingway's *The Sun Also Rises*. It was an extraordinary review, the kind every writer dreams about, and a strange twist of fate which only came about because the regular reviewer, Orville Prescott (who would have hated the book) was on vacation.

The Sunday *Times* was not as enthusiastic. The review by David Dempsey was entitled "In Pursuit of 'Kicks'" and spoke of "depravity," "bohemianism," and "the hot pursuit of pleasure."

When my father read this review he forgot how much he had liked Jack. He forgot the kind, serious, if eccentric, man who had charmed and disarmed him on that bourbon-soaked evening in Scarsdale. When he started bad-mouthing Jack, I defended him. I remember telling Dad that Jack's only sins were sins of omission and sins against himself.

Warren circa 1953

Chapter Seven
SCIENCE AND RELIGION

The average citizen tends to think that science has destroyed the element of faith in religion; instead, he should realize that science is itself founded on faith.[19]

Warren Weaver

From time to time Dad sent me copies of articles he had written that he was particularly pleased with, or thought might interest me. It pains me to say it, but I remember that often I was too busy with my life – looking for love in all the wrong places, hanging out with my friends, and yes, making a living – to pay much attention to his reprints.

And I don't think I read any of his books – at least, not in their entirety – until after he died. Of course, some of them were – are – way too technical for me; but others are not. I think I still carried some lingering resentment for the way he had monopolized the conversation at the dinner table. He was a powerful presence, even after I left home, and for a while I continued to tune him out. This was a shame, because beneath our differences we had so much in common. For example, we shared an interest in philosophy.

For as long as I can remember I have been fascinated by ideas like fatalism, time and eternity, the origin of the universe, the purpose of life, and so on. The apparent contradiction between free will and determinism I disposed of in high school when I concluded that both were true. To deny determinism was to deny causation. Next problem!

At Oberlin Professor Lucius Garvin gave me an A+ in his introductory course in Philosophy. He told me it was the only time he'd ever done that. In those days I could understand, remember, and cough up the details of vast systems of abstract thought I'd have a hard time just grasping today.

My father delved deeply into the philosophy of science. He had a lifelong interest – an obsession, really – in the relationship between science and religion. All his life he tried to reconcile the Christian religion in which he was raised with the rational materialist world view that dominates modern science.

Religion for him was Christianity. He respected other faiths, but they were not for him. The oriental religions seemed too passive to appeal to him with his energetic attack on life. He had a sort of romantic attachment to struggle. I think he would have been a great admirer of the fourteenth Dalai Lama, with his deep respect for science, but the concept of nirvana – the whole idea of aspiring to what he saw as a state of blissful indifference – was alien to him.

In an article entitled "Peace of Mind" that appeared in the *Saturday Review* in 1954, he begins by confessing:

This has been a long time coming to a head. Some years ago I started to read Joshua Liebman's book "Peace of Mind." It irritated me so that I abandoned it after two or three chapters. I was so upset that the mere mention of peace of mind would stimulate my adrenals and cause me to flail about with all the logical and ethical quarterstaves at my disposal.[20]

But the emotional comfort he received from Christianity was a very different matter. Although he had long since abandoned the fundamentalism of his Lutheran ancestors with its literal interpretation of the Bible, he did believe that "at the central core of religious thought there are certain basic principles – the Ten Commandments, the Sermon on the Mount – which give every evidence of being of permanent and immutable value."

Have you taken a look at the Ten Commandments lately? Numbers five through ten look good, but number one is a recipe for religious intolerance, number two forbids all religious art, number three restricts freedom of speech, and number four prohibits all work on Sunday. Those four strike me as pretty dated. The Sermon on the Mount, on the other hand, is still as fine a code of conduct as ever.

Dad is on firmer footing when he asserts, in "Science and Faith" (1954), that science, like religion, is ultimately based on faith:

Faith in the regularity of nature, in the inherent reasonableness of natural phenomena, in the discoverability of scientific laws, and, still more basically, in the goodness of knowledge: these are elements of faith which the scientist does not often stop to express but without which he could not proceed.

He concludes that science and religion need not be in conflict:

... if you are prepared to base your life not on imposed authority but on faith – faith in the goodness of God and in the beauty and order of his [*sic*] universe, faith in the power of the minds with which God has endowed us, and faith that it is proper for these minds to go on discovering more and more of the beauty of his universe – then you need not be troubled. Based alike upon faith, but dealing with quite different and non-contradictory aspects of man's total experience, science and religion, as they can be understood today, need no longer quarrel.[21]

I'm not sure these remarks would make much of an impression on Richard Dawkins or the late Christopher Hitchens, but I have a feeling Einstein would be largely in agreement. Wasn't it he who said, "Science without religion is lame. Religion without science is blind."? In any case, there was once a level of civility that characterized the debate between these great realms of thought that has been lost, and I think we are all the poorer for it.

In the much reprinted "Can a Scientist Believe in God?" (1955) my father, by then chairman of the board of the American Association for the Advancement of Science, points out that scientists are especially good at believing in the "unseeable." Quantum physics tells us that the solid reality of the physical world is an illusion. Physics, for all intents and purposes, has become metaphysics.

He accepts the idea of God for several reasons, the first of which is that

... in the total history of man, there has been a most impressive amount of general agreement about the existence (if not the details) of "God." This agreement is not so logically precise as the agreements about electrons; but far, far more people believe and have believed in God than believe – or have ever believed – in electrons.[22]

This respect for ideas that a great many people have believed in throughout history is central to my father's thinking. He was no elitist snob in an ivory tower. He was essentially a teacher, and like all great teachers, he was a good listener. He wanted to know what ordinary people believed. "Can A Scientist Believe in God?" has been translated into nine languages.[23]

Dad's final statement on the relationship between science and religion occurs in the last chapter of his autobiography, *Scene of Change*, entitled "Science, Contradiction, and Religion." In it he makes use of the principle of complementarity that was introduced by his friend Niels Bohr in 1928.

He begins by reviewing the revolution in physics that took place in the early twentieth century. It all began with Max Planck's idea that "energy is not continuously divisible, but that it always occurs in discrete packets of a precisely specified minimum size" – in other words, particles. But Louis Victor de Broglie and Erwin Schrödinger were talking about quantum theory in terms of continuous waves. It soon became evident that neither concept could be discarded, for

. . . such a basic entity as a photon of light had to be treated under some circumstances as if it were in fact wave-like, but under other circumstances as though it were particle-like. The same embarrassing duality was soon found to apply to electrons.

At around the same time, Werner Heisenberg was developing his theory of uncertainty, which proved that it was "impossible experimentally to obtain precise information on *both* the position and the velocity of a particle." This happened because

. . . observation necessarily affected the thing observed. This effect is ordinarily of no importance, being negligibly small when the thing under observation is of substantial size and mass, but is critically important when observing elementary particles.

To put it roughly, if you bounce off enough energy to tell you precisely where the particle *was*, the rebound imparts so much unknown velocity to the particle that you don't know much about how it is moving, and hence about where it now is. This joint uncertainty was not only characteristic of the pair of quantities' position and velocity, but applied equally to various other pairs of quantities that entered into physical theory.

These ideas were so exciting to Niels Bohr that "he pursued them with what Heisenberg described as 'almost terrifying relentlessness.'" In the end, Bohr concluded that

. . . the information obtained about an object by using one set of experimental conditions of observation should not be expected to be the same as, or necessarily consistent with, the information obtained when using a different set of observational procedures. If the second set of observational conditions excludes the first set, then the information obtained by using one set must be viewed as *complementary* to the information obtained by using the other observational procedure. However contradictory the two sets of information may appear to be, they must be accepted as equally valid.

In turning to his ideas about religion, my father is "sustained and liberated" by the concept of complementarity.

His brand of Christianity is reminiscent of that of Thomas Jefferson, who cut and pasted his copy of the New Testament to eliminate all references to the supernatural. Dad had no use for what he called "dogma," which included everything from the Immaculate Conception to the Resurrection. He rejected Virgin birth and the divinity of Christ but, as we have seen, he had a profound reverence for the teachings of Jesus. When he was in church my father refused to recite the Apostles' Creed, for it contained

... sentences which begin with the affirmation, "I believe..."; and in fact I do not so believe.... The only Trinity that I understand is the trinity of God, my brother, and me.... For those earnest souls who seem to think that God dictated the Bible in English, complete with punctuation, to the committee of churchmen selected by King James in 1611, I have sympathy but not much comprehension.

What was left after all these disclaimers was a twofold concept of God: an abstract notion of God as the moral purpose of the universe ... the author of the grand design, ultimately responsible for its intricate beauty and for our evolving capacity to recognize the lovely unity that pervades all the apparent diversity and, when in trouble or when moved by one of the familiar hymns of his childhood, a personal sense of God as

... the ever dependable friend, the loving, comforting, and protecting Father. If the two concepts ... seem inconsistent or contradictory, then I repeat that they arise under mutually exclusive circumstances and can strictly be viewed as complementary.

Similarly, the apparent contradictions within science (wave vs. particle) as well as the apparently conflicting claims of science and religion he accepts as "formally justified in terms of the principle of complementarity."

These chapters on my father's ideas do not begin to suggest the variety of subjects on which he could speak and write with authority. His published works ranged over every aspect of science from the most technical to the most practical, general, or philosophical.

To take them in no particular order, he wrote about language, translation, communication, and information theory; Lewis Carroll as author and as mathematician, and *Alice in Wonderland* from the point of view of adoring fan and professional bibliographer; philanthropy in general and the work of the Rockefeller Foundation in particular, especially in the fields of agriculture and medicine; censorship and freedom of speech; the genetic effects of atomic radiation; the control and reduction of armaments; education and the teaching of science; science and the citizen; what a moon ticket would buy (he thought we should stay home and spend the money on things needed on Earth); moral problems posed by modern science; science and imagination; basic research and the common good; and on and on.

He was a man of wide-ranging interests, persistent curiosity, a gift for communicating his many enthusiasms to others – and a mission to explain science to everyone, in words that everyone could understand.

Perhaps the most eloquent expression of his ideas and feelings about science, the spirit, and life itself is to be found in a prayer he wrote in the nineteen-forties, to be read to a group of high school seniors in the Sunday school of the First Congregational Church in Scarsdale:

A Scientist's Prayer

Our Father in heaven, we thank Thee for the orderly beauty of the universe in which Thou hast set us. We thank Thee for the serene light of the evening planet, for the majestic pageant of the stars, for the misty loveliness of the whirling nebulae, – and especially we thank Thee for the dependable precision of the laws which govern all their motions. We thank Thee for this assurance, on so grand a scale, that we live in a universe which is disciplined and reliable. The more we discover of the history and laws of Thy heavens, O Father, the more we appreciate and understand the poet who speaks of the music of Thy heavenly spheres.

We thank Thee, O Father, for the orderly beauty of the nature which more intimately surrounds us, – for the delicate symmetry of the snowflake; for the intriguing diversity of living forms; for the manifold patterns, both lovely and useful, into which matter is organized; for the ever-fresh and vital urge of spring; for the spark of life within each cell.

We thank Thee, O Father, that it is permitted that we learn so much about our own bodies, so much about how we came to be as we are, so much about our long, encouraging climb up from the sea and out of the mud and away from the jungle. For this long climb carries with it the inspiring and stimulating hope that we may climb still higher, that we may some day indeed be fashioned in Thine own image.

We thank Thee, O Father, that we live in a world, a nature, a universe that is ever progressing, rather than static. We thank Thee that the laws which describe so many aspects of this progress are discoverable to us. We thank Thee that we need not live in a dark, chaotic, unpredictable, and frightening world of superstition and fear; but that we have been blessed with a capacity to analyze and understand a nature which is orderly and logical.

But we pray, O Father, that our little knowledge not make us arrogant or presumptuous. Help us to remember that our knowledge is, at best, superficial and limited. We discover the laws of matter, – but we do not know what matter is. We study the steps by which evolution has proceeded, – but we do not understand the essential urge that spurs evolution on. We analyze the behavior of living things, – but we do not know what life is. Help us to remember that deep within all these problems there remains a central core which our science does not penetrate. And help us humbly to recognize that this central core, this living essence, this one central and remaining mystery, is in Thy divine keeping. Amen.

Helen in Greece circa 1960's

Chapter Eight
THE LIGHT OF GREECE

There are two ways to live. One is as though nothing is a miracle.
The other is as though everything is a miracle.

Albert Einstein

As my thirtieth birthday approached I decided to quit my job at Farrar, Straus and go to Europe. I was afraid that if I didn't do something I would spend the rest of my life reading unsolicited manuscripts and writing rejection letters to nut cases in California who thought they could write.

I sailed to England on the *Liberté*, spent time in London and Paris, and then worked my way down to Florence, where an Oberlin friend had a villa overlooking the town. I arrived in Fiesole on my birthday and was given a champagne party.

In Florence I took up with an American artist I'll call Andy who drank, didn't do much painting, and had a reputation for being a wild man. I was attracted to his good looks (he was a dead ringer for Steve McQueen) and his bitterness, which I thought I could heal. We lived off the checks Dad was sending me for Italian lessons. Andy called him Generous Warren.

We hitchhiked to Brindisi and sailed to Greece, sneaking into first class and conning food from the cook. Andy was the archetypical con man. I rationalized that I spent less money supporting him than I would have on my own. He taught me to live on nothing at all. We spent two summers on Greek islands: the summer of 1961 on Cos, and the summer of 1962 on Lesvos.

While we were living on Cos that first summer something happened that permanently altered my perception of reality. It may sound trivial; but to me it was a revelation.

Every day we would pick up lunch for a few drachmas and have a picnic at our favorite spot on the beach. And every day a colony of ants would collect our crumbs and haul them back to a nearby ant hill. Sometimes we would smoke a joint and become absorbed in watching the activities of the ants. We would drop stones on the ant hill, blocking the entrance, and watch in fascination as the ants scurried around, creating another entrance in seconds.

One day, after one of our bombings, we noticed an ant that had been wounded. Another ant was struggling to drag the wounded ant back to the hill, but she wasn't strong enough. The helper ant went back to the hill, disappeared for a moment, and re-emerged with a third ant. Together they returned to the wounded ant and, one on either end, as if holding a litter, they carefully dragged their comrade back to the hill.

They might as well have had little red crosses on their caps.

There were no more bombings after that.

When I told my friend Sarvananda Bluestone this story he remarked that maybe they were dragging the wounded ant back to the hill to be eaten. Maybe so. I just remember that we were ashamed. We felt that the ants were better than we were. We were treating them like things, but they weren't things. They had intelligence, they communicated with one another, and they took care of their own.

This experience opened my mind to the idea that everything has consciousness. I called it my Ant Illumination.[24]

In April 1962 we arrived on Lesvos and for a few drachmas a month we found a house in a small fishing village called Molivos. Shortly after we moved in I landed my first translating job through the kind offices of Richard Howard back in New York. The book, *Je Crois en Dieu* (I Believe in God), was a meditation on the Apostles' Creed drawn from the theological writings of the Catholic poet Paul Claudel.

I flew to Athens to buy a typewriter and a French-English dictionary in case I got stuck. I sent lists of questions to my mother in Connecticut, such as the exact wording of quotations from the Catholic edition of the Bible. She would look them up and send me back the answers.

With my Brownie camera I took photographs of the island and sent the undeveloped film to my father, who sent me back prints, carefully labeled and dated.

When we first arrived, it was windy and cold. The wind – it had a name which I've forgotten – was so powerful it wrapped the clothes around the clothesline. But after a few days the sun came out, the sky turned the most amazing shade of blue I had ever seen, and the sea became turquoise. The houses were white. The *streets* were white! The streets of Molivos were white-washed. Suddenly, there were flowers everywhere. Never in my life had I been surrounded by such natural beauty.

The donkeys wore turquoise worry beads around their necks to ward off the evil eye. The water we drank was delicious, and a source of pride among the people. The women I met walking down the white-washed streets bore baskets of unbaked bread dough on their heads on their way to the communal oven. They smiled at me and said "*Chairete*," which I found out means "Rejoice."

For the first time in my life I was living in a place without a constant supply of electricity or running water. These were available for a few hours during the day, so we had to make sure our earthernware vessels were filled. I bathed by heating up water on a kerosene stove and squatting on our porch in the sun. I was amazed at how clean I could get with just a little soap and a gallon of warm water. I rinsed off by pouring cold water out of the amphora over my head, feeling like something out of the fifth century B.C.

I went to bed with the sun and got up when those fiery Aegean rays came blasting through the shutters of our bedroom. I timed my daily activities to its cycle. As I sat at an old wooden school desk working on my translation, shepherds grazed their flocks on the hill outside my window. One Sunday morning a baby goat was born on our lawn. I watched the shepherd help the mother and the wet, leggy creature take its first steps.

On that island everything conspired to heal my insularity and reconnect me to the rest of the creation. I had considered myself an atheist for years, but somewhere along the line that summer I realized that the title of the book I was working on – if not its very Catholic contents – was also true for me. It was not a sudden awakening: it was a gradual remembering. I realized that I had believed in God all along, I had simply forgotten that I did. My God – if I even used that name – was not the God of the Old Testament or even of the New. It was the God of nature, of the sun and the water. The God of everything and in everything. It was the light of Greece.

It was the best of times and the worst of times: this joyous awakening took place while I was trapped in what had become an abusive relationship. Andy and I were no longer lovers, or even friends. I was very conscious of the fact that I was living with a man who didn't make love to me, took my money, and was not above hitting me when he was drunk. I started drinking myself because it made it easier to live with him.

After I mailed the last of my translation to Holt, Rinehart and Winston I came down with hepatitis. Andy took care of me for a while, then got bored. He would go off and leave me for days without food. I remember that I used to picture my father coming out of the sky in a helicopter to scoop me up and take me home. I was afraid. But I knew that I had got myself into this mess and that somehow I would get myself out of it. I vowed that if and when I ever got home I would get professional help.

With the help of friends I managed to get on a boat. I was so intimidated by Andy that I pretended I was only going to New York for a few weeks to sell some of his paintings. I left all my possessions behind, taking nothing but my typewriter and a cat named Petra whom we had adopted in the spring.

Crossing the Atlantic in November, the weather was terrible and everyone was seasick except me. I had to eat, so I stood on the deck holding onto the mast with one hand and a book with the other. I survived. So did Petra.

I spent several months recuperating in my parents' home in Connecticut. I wrote a poem "To A Cat Who Also Escaped By A Hair." As soon as I was strong enough to move back to my Village apartment, I started looking for a shrink. I painted the floor of my apartment white to remind me of the whitewashed streets of Greece.

I had hit bottom and I was ready to change. I had picked someone I thought was wild and free, who would not smother me as my parents had, and I had ended up in a kind of prison, praying for Daddy to come and rescue me. Once again I had been drawn to someone who couldn't love me. Somewhere inside, I knew now, I didn't believe I deserved to be loved. That's what had to change.

And it did change. This time I found a real analyst, one who listened and who actually answered my questions with something other than "Why do you ask?" In 1965 I came as close to getting married again as I ever got in this lifetime. That affair turned into a friendship instead, but not before it put up a milestone in my life. At age thirty-four, in the middle of the sixties, I had my own personal sexual revolution.

In some mysterious way I still don't understand, my sexual awakening in the mid-sixties was inseparable from the spiritual awakening that had started in Greece but continued to unfold after I became orgasmic. Perhaps it had to do with my expanded capacity to feel. It all felt like part of the same miracle.

Now that I had been to Europe and spent two summers on beaches and had a taste of the freelance life, there was no way I was going back to a nine-to-five job. I decided to try and make my living as a translator. This made my father very nervous. I think he was afraid I would end up a bum on the Bowery. But I had a very powerful ally.

Dick Howard farmed out some of his translating jobs to me, and in overseeing and improving my work, he taught me my craft. Eventually he turned the books he didn't have time for over to me, so I had a new career. It was hard work, tedious and solitary, and the books I got at first were not that interesting. But I could set my own schedule and work in my own space, and that made all the difference.

Best of all, I fell in with a group of brilliant, lively artists and poets, mostly gay, and started having a really good time.

In 1969 I slept with a sculptor who looked like a cross between Bob Dylan and Rudolf Nureyev, and got the clap. A doctor in the Village prescribed massive doses of oral penicillin that completely defoliated my digestive tract, causing a case of antibiotic colitis that went on for four months – and didn't cure the clap. I became so weak that I had to go and stay with my parents in Connecticut. I didn't tell them I had the clap.

At last I was well enough to go back to the city. I had started meditating, and now I was ready to make some changes in my "lifestyle." I took up yoga. I started eating brown rice and other healthy food. I took up astrology.

Warren Weaver at home in his Connecticut study 1960

Chapter Nine

A MAD TEA PARTY

> In a Wonderland they lie,
> Dreaming as the days go by,
> Dreaming as the summers die;
> Ever drifting down the stream –
> Lingering in the golden gleam –
> Life, what is it but a dream?
>
> Lewis Carroll, *Through the Looking-Glass*

It must have been around 1900 that a shy and lonely boy in a small town in Wisconsin was introduced – perhaps by his mother's voice? – to *Alice in Wonderland*. Alice was my father's first love and he remained true to her all his life. His first copy of *Alice* – marked "No.1" on the flyleaf with a child's wobbly stamp set – is a family treasure. When he sold his Carroll collection to the University of Texas at Austin in 1969, he kept that copy back. I'm not sure if this was because he felt it had no value – it was pretty dog-eared – or because he couldn't bear to part with it.

In his book about the translations of *Alice* Dad writes:

I think we ought not to try to explain Alice; we should just be thankful that we have her. Innocence, a tiny but truly patrician courage, a steady determination to get things straight, a movingly sympathetic attitude toward all around her, a demure decency that is as appealing as it is rare – all these belong to Alice. And all these are set for us in a matrix of the most deliciously irresponsible humor – a humor that is made all the more enjoyable by the fact that simple Alice, sweet Alice, enchanting Alice, herself never quite catches on.[25]

On his travels for the Rockefeller Foundation my father always found time to seek out secondhand bookstores and look for foreign editions of *Alice*. He had friends all over the world who kept their eye out for books to add to his collection. A nun in Africa helped him track down a Swahili edition. Fellow Carroll collectors in England swapped extra copies with him. There was even a group in Istanbul called "The Turkish Alice Warren Weaver Society."

On trips to India my father spent many hours – and rupees for cabs – trying to run down copies of various editions in Hindi:

In a small bookshop near the Kashmir Gate in Delhi, I found the third edition of one translation. When I explained that I would be very glad to buy the first and second editions, the mystified shopkeeper explained that the story was "just the same" in all editions. After a short lecture on the irrationality of book collecting, I told him I would gladly pay ten or twenty times as much for the first or second edition (not a rash bid, for the third edition had cost twenty-eight cents). Whereupon he gently took me by the arm and escorted me out of his shop. I am confident he thought he was dealing with a madman.

Needless to say, Dad hated the Freudian interpretations of *Alice in Wonderland*. I don't know if he ever tried reading William Empson's "The Child as Swain" in his *Some Versions of Pastoral* (sample: "She is a father in getting down the hole, a foetus at the bottom, and can only be born by becoming a mother and producing her own amniotic fluid."), but I'm sure that if he did, he would have flung the book across the room in disgust.

In 1961 a literary club in Cleveland asked my father to give a talk about his Carroll collection. He decided to talk about the translations of *Alice*. At my suggestion, he called his talk "Alice in the Tower of Babel." I thought that would make a good title for the book he planned to write, but in the end the book became *Alice in Many Tongues*.

Alice in Wonderland presents a special challenge to the translator because of the numerous puns, parodies, songs, jokes, and plays on words. Dad was curious to see how the different translators had handled these challenges. But since he was only fluent in English and German, he rounded up a group of bilingual friends to help him study and evaluate some of the translations.

At that time Dad had 160 foreign editions of *Alice* in forty-two languages. (Eventually he acquired over 250 separate translations of *Alice* into foreign languages, by far the largest number of any other collection.) He chose fourteen of the earliest or most interesting translations, those in German, French, Swedish, Italian, Danish, Russian, Japanese, Chinese, Hebrew, Hungarian, Spanish, Polish, Pidgin, and Swahili.

He selected a passage in Chapter VII, "A Mad Tea-Party," a wild dream sequence that presented an unusual number of obstacles to the translator. He had photostats made of that passage in each of the target languages and distributed them to his team of translators.

Dad asked each member of his team to take the translated version of that passage in their language and, without looking at Carroll's original text, translate it literally back into English so that he could look over the original translator's shoulder, as it were, observe their process, and judge their degree of success.

He classified the problems *Alice* presented to the translator into five groups: verses, mostly parodies of well-known English poems; puns; nonsense words; jokes involving logic; and what Dad called "twists of meaning." The passage in question contained one parody of an English verse, based on the familiar "Twinkle, twinkle, little star;" three puns; one nonsense word; one logical joke; and eight twists of meaning.

To take just one example, the parodied verse, Dad saw three ways a translator could approach the problem: 1) find a verse that would be familiar to readers of the target language and write a parody of it in a style as close as possible to Carroll's; 2) translate the English verse on a more or less literal basis – obviously a less satisfactory solution, but one that would be necessary in cases where there was little or no children's literature to choose from, as was true of Japanese and Chinese as well as Swahili and Pidgin; or 3) since the parody was always humorous ("Twinkle, twinkle, little bat! / How I wonder what you're at"), create an entirely new nonsense verse in the target language.

The first and best procedure was followed by the Danish, French, German, Hebrew, Hungarian, and Russian translators; the second, by the Chinese, Japanese, Pidgin, Spanish, and Swahili translators; and the third by the Italian, Polish, and Swedish translators.

Of the best solutions, Dad found the Russian version, done in 1923 by a translator identified on the title page as "V. Sirin," to be "especially clever and sensitive." This was not surprising given that "V. Sirin" turned out to be the pen name of the young Vladimir Nabokov – who, legend has it, was paid a whopping five dollars for translating the entire book.

The friend who agreed to work on the Chinese translation was Dr. Yuen Ren Chao, a professor of oriental languages at the University of California at Berkeley. Dr. Chao was actually the translator of the first Chinese edition of *Alice in Wonderland* in 1922. Dad described him as "very possibly the most perfectly bilingual Chinese-English scholar who has ever lived." (Try to remember his name, because he's going to be important a little later in this book.)

Among the several Chinese editions one, a student's edition of 1927, was banned by General Ho Chien, Governor General of the Province of Hunan. This gentleman issued an edict forbidding its use in schools on the grounds that it was degrading for human beings to converse with animals!

While Dad was writing *Alice in Many Tongues* I was just getting started on my career as a professional literary translator from the French. Some time in the sixties I translated a book by the French sociologist Raymond Aron that was part of a series edited by Ruth Nanda Anshen. Dr. Anshen invited me to a black tie dinner at which I had the honor of meeting Margaret Mead.

I knew that Mead knew my father, so I went up to her after dinner and introduced myself. When she heard that I was Warren Weaver's daughter she looked pensive for a moment, as if fishing in her memory, then banged her staff on the floor and announced, "Warren Weaver! Your father was depressed in the thirties, so I translated the Mad Tea-Party into Pidgin to cheer him up!"

In a footnote in *Alice in Many Tongues* Dad quotes a letter he received from Dr. Mead reporting on the progress of her Pidgin translation that read as follows:

Master!

This fellow man em i stop along New Guinea, em i work i'm now. I think pretty soon paper he come up. He no long time too much. That's all, along one fellow moon more, me go finish along bush.

M. M.

Thirty years later, Margaret Mead obliged her old friend by translating the passage from the Mad Tea-Party back into English.

Although I never had to deal with any text that had the sheer number of challenges facing the translators of Lewis Carroll, later I did get to take a crack at translating poetry. In translating Philippe Ariès' magisterial study of attitudes toward death in the Christian West which Knopf published as *The Hour of Our Death*, I had the pleasure of working on a favorite poem I had encountered in high school: Lamartine's immortal *l'Immortalité*. Against all the odds, I was able to come up with rhymes that did not sound far-fetched. That is grace! There was poetry in the Artaud *Selected Writings* I translated for Farrar, Straus too, but since it was free verse, it was less of a challenge.

So I learned firsthand what Dad suspected: that when the translator is faced with poetry – as opposed to technical language, at the other end of the scale of difficulty – you do not translate words, but images or ideas. In essence, you create a new poem.

Dad concluded that the most successful translations owed their excellence more to the skill of the translator than to any inherent qualities of the target language. While skill is certainly required, I would respectfully disagree. I always felt grateful that my job was translating from French, a language that is relatively poor in its number of words, into English, that vast storehouse which has borrowed freely from so many different sources. I suspect that to translate from English into French would be more difficult, even for someone far more gifted than me.

At the end of *Alice in Many Tongues*, Dad commented on how the various translators handled the proper names:

The Dormouse is replaced by the equally sleepy Woodchuck in both German and Polish, is called Hazel Mouse in Swedish, Lemur in Swahili, Marmot in Hungarian, Mole in Spanish, and is simply given the name Sleepyhead in [Nabokov's] Russian.

The March Hare is called exactly that in Chinese, Italian, Pidgin, Polish, and Russian; in Danish, French, and Swedish is called simply Hare; in German, the Silly Hare (*faselhase*); in Hebrew, the Hare of Nissan (Nissan is the Hebrew name for the first month of spring, corresponding to March); in Japanese, the Wild Rabbit; in Swahili, the Tortoise; and in Hungarian, the Easter Rabbit!

My father had a special fondness for the White Rabbit, perhaps because it was he who got the story going by leading Alice to the rabbit-hole. Some time after he and Mother moved into the house they built in New Milford, Connecticut Dad had a local artist named Philip Kappel do an etching of the house. He was pleased with the result, but the large white space at the bottom of the drawing told him that something was missing. The artist was not at all offended and happy to oblige when Dad asked him if he couldn't please add, in the lower left-hand corner, a tiny version of the Tennial drawing of the White Rabbit.

Warren Weaver 1962 (Life Magazine)

Chapter Ten
DAD'S HUMOR

Two things are infinite: the universe and human stupidity;
and I'm not sure about the universe.

Albert Einstein

I've always believed that a sense of humor is a sign of intelligence. It seems to me that no first-class mind is without one.

Dad was a great story teller. As a child I enjoyed it when we had guests and he made people laugh. I had certain favorite stories that I didn't mind hearing over and over. Dad was fond of that popular limerick of the 1920s:

I don't like the family Stein:
There is Gert, there is Ep, and there's Ein;
Gert's poems are bunk,
Ep's statues are punk,
And nobody understands Ein.

Then there's the one about turtles and the universe which, like the monkeys at the typewriters, has profound philosophical implications:

A college freshman elects a course in philosophy, and one day the professor says: "Today I'd like to take up cosmology – theories of the nature of the universe. There are many cosmologies, ancient and modern. A cosmology of which I am particularly fond is an ancient Indian one, in which the world rests on the backs of three huge elephants, which in turn stand on the shell of an enormous turtle."

At the end of the term the student approaches the professor. "This is a wonderful course, and I am going to go right on with it. But you know, one thing you said bothers me. It's that business about the world resting on the backs of three huge elephants standing on the shell of an enormous turtle. What is the turtle standing on?"

The professor responds: "That's a good question. As a matter of fact, it is standing on the back of *another* enormous turtle."

At the end of the next term the student says: "But I don't get it. What is *that* turtle standing on?" and receives the same answer. He decides to major in philosophy, and whenever he gets a chance he repeats the question, with the same result. Finally he repeats it as he is getting his Ph. D. with the same professor. At that point the professor puts his hand on the student's shoulder in a fatherly manner and says: "Charles, it's turtles *all the way*."[26]

On the whole question of evolution vs. the account of creation in the Bible, my father had this to say:

I'm not going to waste any time discussing whether the world was made in seven days or whether man is descended from monkeys. As we learn more about the language of the great apes I suspect we will discover that, particularly over the last half century, the monkeys have been bitterly protesting any suggestion that they are responsible for us. The two recent political conventions [April 1965] must convince even the most tolerant and kindly chimpanzees that any interrelationship with man must be vigorously denied in every part of the jungle.[27]

Some time in the sixties, Dad's politics underwent an unexpected turn to the left. In the summer of 1968, in a letter to a childhood friend, he wrote:

Our son, who now has the position of national political reporter on The New York Times . . . indicated that he is now almost completely convinced that Nixon will be the Republican candidate and Humphrey the Democratic. This leaves me with a choice which I would prefer not to face.
    Years ago Mary's mother (who was a very staunch Kansas Republican) told me about a friend of hers who said that he was so devoted to his party that he would vote for a yellow dog if it ran as a Republican. Then it turned out that he actually voted for the Democratic candidate in that particular election and when reminded of his earlier statement he said, "Yes, I will vote for a yellow dog, but lower than that I will not go." This pretty well sums up my feeling about Nixon.[28]

A bon vivant who counted among his favorite things in life fine wine, an Old Fashioned made with a really good bourbon, salted peanuts, and Limburger cheese the stinkier the better. It was one of the ingredients of a boiled dressing concoction he called Weaver's Dynamite, which was not ready to eat until fumes rose from the jar upon opening. Dad was not into health food. One of his favorite meals was roast duck followed by pecan pie. He said that every time he had that particular combination he was always sick, and "It was *always* worth it."

My Woodstock friends insist that this chapter would not be complete without mention of my father's recipe for eggplant. Dad was not good about eating his vegetables. He had a particular aversion to eggplant. His recipe for eggplant:

Wash well. Place on a cutting board. With a sharp knife, cut the eggplant in very thin slices. Then throw all the slices away.

What else? I remember that one year a few days before Christmas Dad was shopping on the first floor of Macy's department store on 34th Street in the city. The store was very crowded and the atmosphere rather hectic. Dad happened to sneeze in proximity to a fur-coated dowager who was obviously a germaphobe, for she promptly registered a vigorous complaint. My father drew himself up and replied, "Madam, you are too delicate for Macy's."

After Dad died I found in his papers a page of "brief but illuminating book reviews." A sample:

Unabridged Dictionary: "Interesting but disconnected."

Bible: "Important if true."

Well, Dad's going to need his sense of humor, because his only daughter has taken up astrology.

Helen in the early 1950's

Chapter Eleven

WHY ASTROLOGY?

The important thing is not to stop questioning. Curiosity has its own reason for existing. One cannot help but be in awe when he contemplates the mysteries of eternity, of life, of the marvelous structure of reality. Never lose a holy curiosity.

Albert Einstein

My father wondered how on earth his very intelligent daughter, who had graduated from Oberlin with honors and was making her living as a literary translator, could possibly give any credence to astrology.

How, indeed? The path from skeptic to believer seems accidental, yet inevitable.

As a child I believed in Santa Claus, the Easter Bunny, and (I guess) God. But above all, I believed in magic.

In my dreams I could fly, and there was a time when I had flying dreams almost every night. Flying was both a way to escape any disagreeable situation on Earth by literally rising above it and a delicious sensation in itself. Of course, there was always the danger of flying too high and not being able to get back down again, but that was a risk I was willing to take. The ideal level was just over the tops of the trees: high enough to be free of the Earth's troubles without quite leaving her gravitational field.

When I was awake, I sometimes doubted my ability to fly; but as soon as I was asleep and dreaming, I'd chuckle at my uncertainty and say to myself, "Of *course* I can fly – what was I worrying about!" and off I'd go.

I believed – or perhaps remembered (sometimes it's the same thing) – that my spirit had flown freely through space before it entered my body, so even when I was awake and confronted with evidence to the contrary, I continued to believe in the possibility of flight.

Not all my dreams were happy. There was a recurrent nightmare in which I had to work my way through an impossibly small tunnel toward the light that I've always assumed was a birth memory. My mother confirmed this theory by telling me that she had been given ether:

"It was wonderful!" she said. "Why, I just went to sleep, and when I woke up, there you were!" (*Wonderful for you*, I thought bitterly. *I did all the work!*)

The world of magic and make believe was my element. My first friends were my stuffed animals. But after I started having human friends the first question I would ask them was, "Do you believe in magic?" And if the answer was "Yes," then I knew I was on solid ground.

Magic, of course, had to give way eventually to Christianity, which meant saying my prayers every night when Mother put me to bed ("Now I lay me down to sleep") and ultimately, the First Congregational Church of Scarsdale, New York. I had to go to Sunday school whether I wanted to or not. And like my brother before me, at a certain age I had to memorize the Sermon on the Mount.

Although she was not what she would refer to condescendingly as "pious" and had no use for the miracles in the Bible, my mother did believe in the ethics of Jesus, and she made both my brother and me memorize Chapters Five through Seven of the gospel according to Saint Matthew. At Mother's service my brother remembered this and remarked that the experience had been of greater benefit to his prose style than to his soul.

On Sunday afternoons when I longed to be outside playing with my friends Mother would sit me down in a little chair next to her as she lay on the chaise longue in her bedroom. Then she would open my copy of the Bible, and we would begin: "And seeing the multitudes, he went up into the mountain: and when he was set, his disciples came unto him: And he opened his mouth, and taught them, saying, Blessed are the poor in spirit: for theirs is the kingdom of heaven. . .."

And the great King James version rolled on, telling us to love our enemies and to pray in secret and that we were the light of the world. And instructing us, impossibly (but in my mother's voice, it almost sounded like a possibility she entertained): "Be ye therefore perfect, even as your Father which is in heaven is perfect."

The magic there was in the words. Mother told my brother and me that we would thank her one day, and so we did.

I hated getting up early on Sunday morning for Sunday school and later, for church as well – two whole hours of organized religion. I especially hated the fact that I had no choice in the matter. But I came from a long line of Sunday school teachers and my parents were pillars of the church; so for me there was no escape.

If Sunday school was boring, church was even worse. When I was little, Mother brought pencil and paper, and I was allowed to draw pictures during the sermon. But when I was a teenager, I was supposed to sit up and pretend to listen.

There were things I did like about church. Like my namesake, Grandma Hemenway, I sang in the choir, and I came to love the music we sang. Even though I've long since stopped calling myself a Christian, the hymns of my childhood still have the power to move me and to take me to that place in my heart where my parents live.

The other thing I liked about church was communion. Unlike the Catholics and the Episcopalians, we only got to have communion once a month, and at first I looked forward to communion Sunday because there was no sermon that day. All that business with the bread and wine was a blessed relief, but there was more to it than that. I had been pressed into service to help out in the Primary Department, so I had been backstage, as it were. I had seen where the communion stuff was kept, and I knew it was just Wonder Bread and Welch's Grape Juice. But when Reverend Boynton said it was the body and blood of Christ, and when you put your own tiny portion into your mouth – took the little white cube off the tray and the tiny sip of juice from one of the individual crystal cups – something funny happened.

I knew from nutrition class that the starch in the bread was turning to sugar in my mouth. This doll's dinner that the minister said was Jesus was going into my body, was becoming part of me. And the same thing was happening, the same chemical reaction was taking place, inside every other mouth, in every other body in the church. The mystery of communion – though I could not have put my finger on it at the time – was not so much that we were becoming part of Jesus as that we were becoming part of one another. That all of us, including Jesus, were made of the same flesh, that we were all one body.

Without understanding why, I loved it. I loved the communion hymns, especially:

Break Thou the bread of life
Dear Lord, to me
As Thou didst break the loaves
Beside the sea
Beyond the sacred page
I seek Thee, Lord
My spirit pants for Thee
O living Word!

But singing hymns and breaking bread do not a Christian make. Pagans have those, too. I don't think I ever really swallowed what my mother referred to as "the Christian myth": resurrection, redemption, and so on, any more than my father did.

    The summer I was thirteen I had an experience that changed my beliefs forever.
    Remember those scientists who got useful information in a dream? It happens. The experience that changed my world view, my beliefs about life and death, was a dream. It was another of my dreams of flying. But unlike my childhood dreams, this time I found the courage to leave the Earth's atmosphere altogether and to travel alone and unaided to the moon. It was a long journey of many stages: enterings and leavings of powerful orbits which helped me to conserve my energy; great downward swoopings, as if I were on an enormous invisible roller coaster or ferris wheel; down and around and up again, as if I were in black water. But I was neither cold nor afraid, but exhilarated with freedom, and also with recognition, for all this was somehow familiar, and I knew the way.
    When at last I reached the moon, I played on her surface for a while, jumping high into the air for the sheer joy of it and coming down softly, over and over again until I was tired. I fell asleep and had a dream within a dream.
    When I awoke, I understood that now that I had reached the moon, I could go anywhere in time or space. For my first choice, I decided to revisit medieval France. I flew back to Earth in an instant; now that I had touched the moon, I could go with the speed of light. I zeroed in on northern Europe and the green hills of France. The French and English armies were camped on

adjacent hills, preparing to do battle. As I hung over the battlefield, I could see every detail of the many-colored tapestry below: the fields and vineyards, the villages with their tiny church spires; the rich caparisons of the horses, the steam rising from their nostrils in the cool dawn air, the sunlight glancing off my own armor, the tired, bearded faces of the men.

After France, I revisited other lifetimes, in England, Germany, and Greece, but alas! The rest is lost.

I awoke from this dream with an indescribable sense of exaltation. I was totally under its spell and lived within its resonance for days. I knew as surely as I knew my own name that this was not "just a dream." I knew that I had been on a journey and that I had learned something important about the nature of reality. The dream had the unmistakable stamp of truth about it, and it left me with the sense that a whole dazzling range of possibilities was within easy reach; that this exalting travel through space and time, this panoramic perspective, was just as human and natural as sleep.

The dream seemed as clearly a memory or re-enactment of the spirit's ability to travel freely through time and space as that other dream, its polar opposite, the claustrophobic nightmare of the tunnel I had to crawl through to reach the light, seemed an unmistakable memory of birth.

Thirteen Moon Dream, as I call it, did not so much change my ideas as remind me of a truth I had already known, but had forgotten. I do not know whether at thirteen I had ever heard of the doctrine of reincarnation; but I am sure that this dream helped to open me to the possibility that all of our living and learning may not necessarily be confined to a single lifetime. I only remember that I wanted to be back there: I wanted to live in that infinite space and freedom, with the option of touching down on Earth whenever I missed the colorful distractions of history. For days I stumbled around Candlewood Lake Club staring with homesick longing at the moon.

History was never my strong suit, and for years I assumed that the scene I had observed from above in France had taken place before the battle of Orléans, which in reality looked nothing like that. And for years, I identified with the Maid of Orléans, most of whose battles were fought against towns. The scene I saw in my dream probably took place much earlier, perhaps at Agincourt. But like every other lunatic on the planet I thought I might have been Joan of Arc. After reading several biographies of Joan I became disenchanted with her. I decided that I would never have trusted that creepy Dauphin, and that I was probably just some soldier who had once ridden into battle in medieval France.

In high school I had a brief surge of interest in Christianity when my friend Eleanor Gates made me read *The Robe* by Lloyd C. Douglas. A novel about the early history of the Church, it tells the story of a young Roman soldier who is present at the crucifixion of Jesus, wins his robe in a gambling game, and is eventually converted to Christianity. It's about a skeptic who becomes a believer, something that has always fascinated me, and I was immediately converted. My Christian period didn't last very long, but while it did, I decided to become a member of the Scarsdale Congregational Church.

The ceremony was a big anticlimax. One Sunday morning a handful of "young people" who had completed the requisite Bible class sat up front in the first pew. The Reverend Boynton came down from the pulpit and made a little speech welcoming us into the flock, and then we each went up and shook his hand. That was it. It was a far cry from the drama of the Passion, the horror of the arena, or the candle-lit catacombs of Rome.

As soon as I got to Oberlin I stopped going to church. I slept late on Sunday mornings and put a sign over my door with the words inscribed at the entrance to the underworld in Dante's *Inferno*: *Lasciate ogni speranza, voi che entrate* ("Abandon all hope, ye who enter here").

I had turned my back on Christianity. And yet my favorite teacher, Andrew Bongiorno, who taught seventeenth-century literature and Dante, was a devout Roman Catholic, and you practically had to be converted to get the full benefit of his courses on *The Divine Comedy* and *Paradise Lost*. I was swept away by the metaphysical poetry of Milton and Donne and Herbert, whose riches Bongiorno laid before us, and at the same time by the sacred music of Bach and Mozart which I sang in the Oberlin Musical Union. During my junior year I was simultaneously immersed in the rolling periods of *Paradise Lost* and the thrilling music of the *Saint Matthew*

Passion. I was saturated with Christian imagery. I loved the drama but rejected the doctrine. I have no idea what I *did* believe in those days, but I'm pretty sure I was an agnostic, if not a downright atheist.

All that changed when I went to Greece.

In Greece I lost the sense of separation from the rest of creation that I had always felt without even knowing it. For the first time in my life, I felt connected to the natural world: to the daily rhythm of Earth and sun. From there it was a short step to feeling that everything was connected. My experience with the ants had shown me that even insects have consciousness. Perhaps the whole world was alive.

In the sixties I was no stranger to mystical states induced by music and marijuana. I started to see things as joined by invisible bonds of affinity, to see the world as full of hidden correspondences. My journals from those days are full of phrases like "the hidden harmonies in the created world;" statements like "the world is one and I am a part of it" and "metaphor reflects the interrelatedness of all life" and "every day is important."

When I began exploring Eastern religions, I found that they took ideas like reincarnation and astrology for granted; and they were far older (and perhaps, wiser) than the Western faiths. In America in the sixties astrology was suddenly in the air: "What's your sign, man?"

I equated Christianity with "duality and guilt," Hinduism and Buddhism with "totality and enjoyment." Christianity was all about living in the future in some bodiless realm; my new loyalty was to the present moment. I was drawn to my friend Allen Ginsberg's ideas about a religion of the body, a Blakean vision of the human form divine. Norman O. Brown's *Life Against Death* with its indictment of Christianity as damaging to the psyche and its call for "polymorphous perversity" became my bible.

Although I suffered from periodic depression I could write about "the unbearable excitement of just being alive."

I began keeping a journal – first as therapy, and then to "find my voice" and to find out what I thought about things because, like everything around me in the sixties – like the times themselves – my ideas were changing.

In 1967 I wrote:

It came to me that what my faith is based on is not any *thing* but the power of faith itself. The fact that faith changes the world is enough to give me faith in the world, for if faith does these things, if it opens doors, then there is something loving and wise at the heart of the system.

This is enough order. An attitude of openness to reality is successful and changes that reality. Communication, and even communion, exist.

Going through these journals many years later, I come upon my first references to astrology. I don't know what I'm talking about – I've obviously never had a birth chart calculated – but I'm drawn to the subject. I read that people with my sun sign, Gemini, love to learn, are interested in ideas and language, and make good teachers, writers, and translators. I recognize myself.

But I learn that the sun sign is only part of the equation; that to get the full benefit of astrology one must have a chart drawn up for the exact time and place of birth. Eventually, I got around to asking my mother what time I was born. She told me that I was "on hand by midnight."

I must have gone to Weiser's – the great esoteric book store that was then down on lower Broadway – and bought some beginning texts, because by October 1968 I'm trying to calculate my Ascendant (the sign of the zodiac coming up over the horizon at the exact moment of birth) and moon sign. I actually did a pretty good job, considering that in those pre-computer days you needed: an ephemeris (a sort of calendar listing the daily positions of the planets), a table of houses (don't ask – has to do with celestial geometry at various latitudes and longitudes), a slide rule, and a basic understanding of logarithms. I came up – correctly, it turned out – with Pisces rising and moon in Leo; but I had only the most rudimentary idea of what it all meant.

I had my first reading with Zoltan Mason, a Hungarian astrologer and teacher with an office on Lexington Avenue right next to Bloomingdale's. I still have a yellowish piece of paper with my typed-up notes from that appointment. There was one false note: "a strong desire for home and children" was not really me, at least not after about age thirty-five; but almost everything else he said was right on target: "a rage to create;" "a streak of masochism;" "mediumistic tendencies;" "unlucky in love;" "youthful appearance:" all true.

He said that the second half of my life would be better than the first, which was under the influence of Saturn. That *is* the story of my life: Greece was the turning point, and life was just starting to get better. He said that I would make a name for myself. That has yet to happen, although I did achieve a certain reputation as a translator, and City Lights did publish my memoir of Kerouac and the fifties, *The Awakener,* to great reviews. He saw that I would have famous friends, or "friends in high places," as he put it: also true.

He also told me in his Hungarian accent, "You fall in lahv with me!" which certainly did not happen. Nevertheless, I was impressed that he had seen so many things about me just from looking at the positions of the planets at my birth. I was very much into self-knowledge in those days, and the whole idea of this complicated language that seemed to provide real insights into character and personality was intriguing.

I wanted to know more about it, but I found most of the books on the subject disappointing. Many of them were poorly written and even the better ones were tough going. Worst of all, I had nobody to talk to about my newfound interest. Most of my contemporaries dismissed the subject as not worth investigating. This actually added to its appeal for me. As someone who had felt like a social outcast in high school, I have always been drawn to unpopular causes.

My friend Marcia Newfield was one of the few exceptions to the prevailing attitude. A wonderful poet who makes her living as a teacher, Marcia is as intellectual as they come, but has an open mind. She suggested that I teach her astrology so I'd have someone to talk to about it. When I said I didn't know enough to teach, she said, "Well, why not study with Zoltan Mason?"

Of course! It was the obvious solution. I signed up for his class.

Weaver: TRANSLATION OF LIGHT

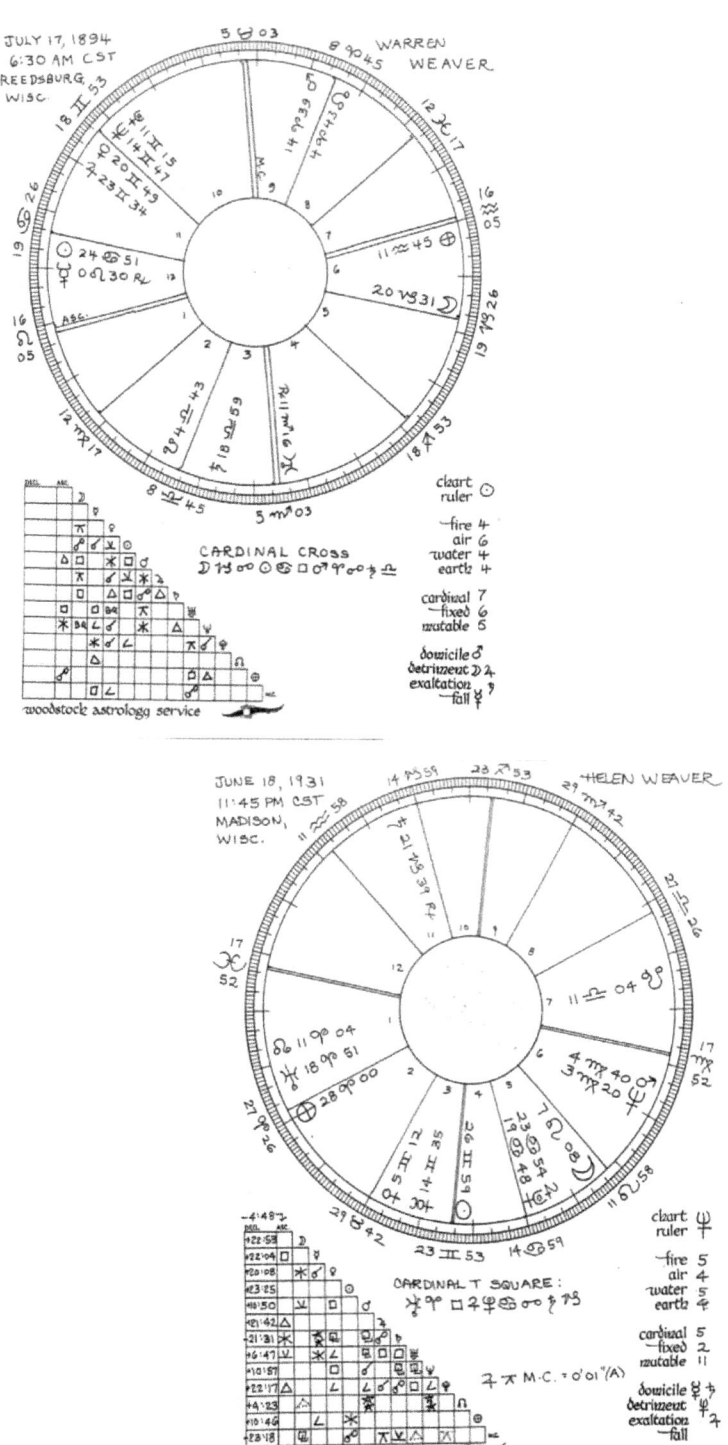

Helen's Astrological Charts for Her and Her Father

Chapter Twelve
LEARNING A NEW LANGUAGE

We are born at a given moment, in a given place, and we have, like the best wines, the quality of the year and the season which witness our birth. Astrology claims no more than this.[29]

Carl Jung

Zoltan Mason was a rather handsome dark-haired man in his early sixties with a voice like Bela Lugosi who was said to be fluent in French, German, Hungarian, and Romanian as well as English. He was not a great teacher. He lectured for the whole hour and when we tried to ask questions he told us to save them for the end of class, by which time we usually forgot what we had wanted to ask.

And yet, looking over my notes from his weekly class, I see that according to his lights – those of the old school, where planets were classified as "benefic" and "malefic" (helpful or harmful) and where free will was not always an option – he actually did give us a good grounding in the basics of traditional astrology.

In his further defense, I came to realize that astrology is as difficult to teach as it is to learn. Although in a real sense it's a language, I found it much harder to learn than French. The trouble is, it's a language of symbols, and the symbols in astrology, unlike most of the words in French, are multivalent. The planets, the signs of the zodiac, and the houses (the twelve pieces of pie into which a chart is divided), which are the building blocks of astrology, each have a whole constellation of possible meanings.

For example, in French, the word "*amour*" means "love." In astrology, the planet Venus can mean love, beauty, art, friendship, favors, money, value, taste, attraction, and/or magnetism, among other things. After a while one can sense that all these meanings have a certain essence in common, but what that essence is cannot be easily reduced to a single word or phrase.

As a consequence, the only way to learn astrology is total immersion; and the best way to bring about this immersion is to supplement one's study of the confusing mass of detail given in the textbooks with a careful examination of the birth charts of the people one knows best. Since self-knowledge is one of the best uses of astrology, it is good to begin with one's own chart. Next, one should concentrate on family members, friends, and well-known historical figures for whom one has accurate birth data (not always that easy to come by).

I was lucky that my mother remembered what time I was born. (People with her sun sign, Virgo, are apt to be pretty good about details.) "On hand by midnight" turned out to be 11:45 PM, and the resulting picture of the solar system at that time on June 18, 1931 in Madison, Wisconsin gave a surprisingly accurate set of clues to my personality, when translated from the language of astrology into English.

Even before my reading with Zoltan Mason I had calculated correctly that the sign my moon was in was Leo, and the description of that sign in the textbooks did ring a bell. The keywords for Leo were leadership, dramatic ability, creativity, and the desire to give and to shine; and on the negative side, pride, self-centeredness, a "star complex," a desire for power and control. Guilty as charged!

Astrologers believe the moon, which stands for the instincts and the unconscious, to be especially prominent in childhood, and as a child I did want to be the center of attention. At Greenacres Elementary School I was a little show-off, constantly volunteering to give oral reports on whatever it was we were studying at the time. And at Candlewood for several years I took the leading role in the play the junior group put on at the end of the summer. I was always nervous, but I always did well, and found I could improvise in character when other kids forgot their lines.

My parents had to keep reminding me that "the world does not revolve around you." Leo knows better!

All this changed when adolescence hit. From about age fourteen on, I was suddenly incapable of standing in front of the class without shaking in my boots. I had a lovely singing voice but refused to consider singing solos. I shunned the limelight. What was going on?

I learned that another important piece of the puzzle is the rising sign: the sign of the zodiac just coming up over the Eastern horizon at the moment of birth. When in my first attempt at casting my chart I came up with Pisces rising, I groaned and hoped I was mistaken, for the books – especially the older ones, which seemed to focus on the negative – made Pisces sound like an "unlucky" sign: shy, dreamy, super-sensitive, physically delicate, susceptible to illness, depression, and addiction; in short, "not really at home on the earth plane."

On the other hand, I read, Pisces people had a natural talent for music, poetry, and drama (again), were compassionate and empathetic, and often drawn to meditation, yoga, and spirituality. When Zoltan confirmed that I did indeed have Pisces rising, I decided I might as well learn to live with it. I wondered if that was why I found myself attracted to so many Pisces people: Chopin, Poe, Kerouac, and many others less well known.

The Ascendant (the exact degree of longitude of the sign rising at birth), I learned in class, represents the physical body and the self one presents to the world, and astrologers believe that it becomes more important as the "native" (the old-fashioned term for the person whose chart it is) gets older. According to Zoltan, the Ascendant was the key to the horoscope, and even more important than the Sun, the Moon, or the planets.

This seems as good a place as any to note that astrologers capitalize Sun, Moon, and Earth, along with the other bodies whose movements they study, so I hope you'll forgive me if I do that from now on. Besides, it has always struck me as interesting that we capitalize the names for the planets Mercury, Venus, Mars, and so on – but not the name of the planet we live on, which we have tended to take for granted (or treat as dirt under our feet). Fortunately, that is beginning to change, and even the *New York Times* now capitalizes Earth.

Zoltan told us that originally the word "horoscope" – from the Greek *hora*, hour, and *skopos*, watcher – referred to the Ascendant only: to search the sky to see what sign is rising. (Ironically, the word "horoscope" comes from the same root as the word "skeptic," which originally meant one who doubts, but looks at the evidence – presumably, *before* making up his or her mind.)

All the bodies that astrology studies are constantly moving: even our Sun moves around the center of our Milky Way galaxy about once every 250 million years. The Moon orbits the

Earth in approximately 28 days, or about a month, and the Earth rotates around the Sun once a year, and on its axis, once a day. These three bodies, along with the planets, move against the backdrop of the zodiac, the slower moving planets – Uranus, Neptune, and Pluto – spending years in a single sign. So the qualities associated with those outer, slow-moving planets correspond to entire generations.

Since the Earth's rotation on its axis is the fastest of all of these motions, it follows that the Ascendant, whose degree changes in a matter of minutes, is the most personal and sensitive spot on the chart. So that 18th degree of Pisces which was appearing above the horizon when I took my first independent breath is who I am, like it or not: a two-edged sword of super-sensitivity to music, poetry, art – and drugs; a weak but flexible body; an overdose of compassion, the urge to take care of some of the neediest cases I come across, and yes, a tendency to be "unlucky in love," as Zoltan perceived. My list of lemons, bad choices, alcoholics, schizophrenics, freeloaders, and homosexuals – the latter were great guys, but hardly husband material – is pretty impressive.

There were other clues in my birth chart to my single state, my failure to marry and have children and a so-called "normal" life. But first please bear with me while I give a little introduction to the basics.

Just as astronomy inherits the names of the planets from ancient Roman mythology, astrology inherits the four elements of ancient Greek philosophy. The twelve signs of the zodiac are divided rather neatly into three each of earth, air, fire, and water. (N.B.: the word "earth" does *not* get capitalized when it refers to an element.) Of these elements, fire and air are considered masculine and active, and earth and water feminine and receptive. The element of earth is associated with matter, the five senses, and security: the earth signs are Taurus, Virgo, and Capricorn. Water is associated with emotion: the water signs are Cancer, Scorpio, and Pisces. Air is associated with communication and ideas: the air signs are Gemini, Libra, and Aquarius. Fire is associated with energy and spirit: the fire signs are Aries, Leo, and Sagittarius.

The twelve signs of the Zodiac are in turn divided into the three energies (which actually correspond to the three forces in physics): cardinal, for outgoing or centrifugal force; fixed, for indrawing or centripetal force; and mutable, for wave motion. The cardinal signs are Aries, Cancer, Libra, and Capricorn: the signs the Sun is in at the beginning of spring, summer, fall, and winter. The fixed signs are Taurus, Leo, Scorpio, and Aquarius: the signs the Sun is in when the season is at its most intense. And the mutable signs are Gemini, Virgo, Sagittarius, and Pisces: the signs the Sun is in when the season is changing.

In my chart my Sun sign, Gemini, is mutable air; my Moon sign, Leo, is fixed fire; and my rising sign, Pisces, is mutable water.

According to Zoltan, a woman with her Moon in a (so called) male sign – especially self-centered Leo – "has lost the instinctual drive for procreation" and "doesn't know the right man." Tell me about it!

With both Sun and Moon in male signs, I never had a strong desire to have children; I always felt there was something else I was supposed to do. Eventually I realized what that was, but it took many years. Zoltan said that a woman with Sun in Gemini "must wait til around forty for realization. Gemini is not a strong sign because it does not go for one thing." He also said that the Sun in the fourth house (which has to do with the home and the end of life) delays realization. My Sun in Gemini is in the fourth house.

Although I started writing poetry in my teens, I didn't become a translator until my thirties. I didn't discover astrology until I was almost forty. I didn't begin to make the transition from translator to writer until my fifties. I didn't start writing my memoirs until my sixties and getting them published until my seventies. We're talking late bloomer!

The best thing Zoltan said about my Sun sign was: "What can you learn from Gemini? To be a human being." From about age seventeen on, that has been my goal.

If all this sounds like "fate," it wasn't. Even an old-school astrologer like Zoltan believed that "character is destiny." I learned that the planets don't force you to do anything; they describe who you are.

There is a common misconception that astrology denies free will. Some time ago, a book reviewer in the *New York Times*[30] defined astrology as "the belief that human lives are ruled by the stars and planets." While some people may believe this, no serious astrologer does. Modern astrologers believe, as Jung did, that a study of the positions of the planets at a person's birth can yield valuable insights into their personality. By casting charts for his patients he was able to zero in on their problems in less time than would have been required without the aid of astrology.

In any case, I soon started on what became a lifelong project (and what is every astrologer's responsibility): collecting accurate birth data for anyone I knew well or found interesting.

Unfortunately, in the case of my parents no birth records with the time of day were available, and both of my grandmothers were deceased. So I set up "solar" charts: charts cast for sunrise on the day of birth. The rationale behind the solar chart is simply that astrology teaches, and astrologers believe, along with Jung, that beginnings are important. Like a photograph taken with an old-fashioned camera with the lens a few degrees out of focus, with a solar chart, you probably won't get a clear image. Unless the person was born at dawn you won't get the right rising sign, and if the Moon changes sign that day, you can't be sure of the sign of the Moon, let alone the degree of the sign. But you will get the planets, and that is better than nothing.[31]

I remember asking my father if he knew what time he was born, and him telling me he had never been interested in knowing this. When I asked him why, he said, "Because I have no use for the information." On the strength of this I decided he must have a practical earth sign on the Ascendant, and I settled on skeptical, intelligent, hard-working Virgo. My German ephemeris told me where his Sun, Moon, and planets were on July 17, 1894, so I was able to set up a tentative working chart.

Warren Weaver was born at the Full Moon, which means that in his birth chart the Sun and Moon are in "opposition," or approximately 180 degrees apart, and (as usually happens) in opposite signs. His Sun sign, Cancer, is the cardinal water sign, and therefore outgoing, feminine, and emotional. Cancer is actually the sign that rules motherhood, among other things, and I've often felt that my father was more like a mother to me: more outwardly affectionate and nurturing than my reserved Virgo mother.

Zoltan thought Cancer was a bad sign for a man unless he was in one of the helping professions like medicine or psychology that required good listening skills. The whole purpose of a philanthropic foundation is to help people. An important part of Dad's job at Rockefeller (and all of his work at the Sloan Foundation) was connected with medical research.

Dad's Moon is in Capricorn, the cardinal earth sign, so outgoing, feminine, and practical. Capricorn is characterized by ambition, discipline, and organization. It also happens to be one of the signs most often associated with science. Louis Pasteur, the world-famous French chemist and biologist, had six out of the ten planets in Capricorn, a rather unbalanced situation which Zoltan said amounted to fanaticism.

Zoltan had this to say of someone like Dad whose Moon is in earth and whose Sun is in water: "Realization will be difficult, as water is feminine and very emotional. If you have this combination you must fall in love with your work."

In his autobiography my father writes: "My years at the Rockefeller Foundation were deeply satisfying to me. I loved my work. Every morning on the commuting train I would go over, with pleasurable anticipation, what was scheduled for that day."

About the Full Moon in general Zoltan said, "In birth [it] is excellent. It does not give instant happiness, but it can give strong motivation."

Later, when I knew a little more about astrology, I discovered that my father's Sun and Moon were part of a relatively rare major configuration called a Cardinal Cross which is often found in the charts of individuals who come from humble beginnings and go on to achieve a great deal. Dad was the son of a struggling druggist in a small town in the Midwest who came to be recognized worldwide for his achievements in science, mathematics, and science administration, among other fields. As we have seen, the mathematics building at New York University, Warren Weaver Hall, is named for him.

The other sign associated with science is Gemini, probably because Gemini people have a kind of universal curiosity and love of knowledge. Dad had four planets in Gemini, one of which, Jupiter, was only three degrees away from a conjunction with my Gemini Sun. Jupiter is associated with growth, abundance, expansion, wisdom, and philosophy. Known to the ancients as "the greater benefic," it was believed to bring good fortune.

The study of relationships between charts is called synastry. I find it particularly compelling because in every case I've known where two people are strongly connected through work, love, or some sort of important learning experience, the mathematical precision (and often the number) of the connections between their charts is astonishing.

Conjunctions between charts (planets occupying the same degree, or close to it) seem to bring people together for some purpose. Dad and I have multiple connections between our charts, too numerous to go into here. Suffice it to say that the Jupiter-Sun conjunction worked both ways: not only is his Jupiter in Gemini conjunct my Sun, but my Jupiter in Cancer is conjunct his Sun.

We brought each other good fortune. Dad signed my copy of *Lady Luck* "For Helen – about the best luck I ever had – Dad;" and I felt the same way about him.

Helen with her Mother and Father at her parents' Connecticut home circa late 1960's

Chapter Thirteen
DIALOGUE

The belief in the effect of the constellations derives in the first place from experience, which is so convincing that it can be denied only by those who have not examined it.[32]

Johannes Kepler

"Horrified" would be too strong a word, but my father was certainly mystified when I took up the serious study of astrology. I'm not sure how he got wind of it, but I can see by the letters we exchanged when he and Mother were in California in February, 1969 that I was definitely out of the closet. (If I was ever in it: perhaps I didn't realize how upsetting and confusing my interest in it would be to my scientist father.)

I was in a state of confusion myself. As is so often the case for those of us who discover astrology relatively late in life – after all, nobody taught it to us in school – I was at a turning point in my life. I desperately wanted to make the leap from translator to writer, but how was I going to make a living in the meantime? Richard Howard thought my poems were publishable and had even offered to help me get them published; but poetry was hardly the answer.

I wrote my parents asking them to consider the possibility of giving me a sort of literary scholarship so I could take a year off from translating and focus on my writing: an experiment. This was especially surprising to my father since the previous year I had written him a long letter holding him and Mother responsible for "overprotecting" me and asking him to discontinue my monthly check. Poor Dad! No wonder he was confused.

As always happened with Dad, after firing off a letter in which he doubted he could afford any such program, he slept on it and figured out a way to give me what I asked for. It was not as easy as I assumed: he was now almost seventy-five and semi-retired.

Dad had been on the board of trustees of the Sloan-Kettering Institute for Cancer Research since 1954 and was chairman of that organization's Committee on Scientific Policy.

After his compulsory retirement from the Rockefeller Foundation in 1959, he had joined the staff of the Sloan Foundation, which did not have a formal retirement age. He had signed on as Vice President, and after travel to the city became too much for him, continued to serve Sloan as a consultant.

In the summer of 1969 my father sold his Lewis Carroll collection to the University of Texas at Austin.[33] I could hardly bear the thought that Dad's collection – acquired with such painstaking pleasure over the years – was to disappear from the glass-covered shelves that had been especially designed for it, and from his life: that he wouldn't be able to add to it, putter over it, and enjoy it in his old age. I remembered seeing those dolls of Alice and the White Rabbit and the Mad Hatter as a child, and my nieces, his grandchildren, remembered them, too. When the collection was being packed up I secretly kept out a tiny wooden oyster that had stood next to the Walrus and the Carpenter. I never regretted the theft: he seemed too small to go to Texas.

The thought that my drain on Dad's resources was in any way responsible for the sale is painful to me now. But perhaps it was, as he said, that he wanted to spare my mother the burden of disposing of the collection after his death.

In any case, he sent me the first of the larger "experimental" monthly checks, and in the very loving letter that accompanied it, he remarked:

Some day you will have to explain to me how you possibly *can* take astrology with any seriousness! I would have supposed that either your intellect or your sense of humor would have excluded that.

I countered with the obvious influence of the moon on terrestrial events, from its well-known effect on the tides to the more controversial but well documented effects of the full moon on human behavior. I told him to ask anyone who works for the police department or a mental hospital.

In October Dad wrote his friend Gerry Piel, editor of the *Scientific American*, suggesting that he get some responsible articles on known terrestrial influences of astronomical events; and is there not some astronomer who could write a really devastating piece about the amount of non-science and conscious or unconscious fraud involved in the present furor for astrology?

This did not happen; but almost a year later the subject was still on Dad's mind. He couldn't leave it alone, any more than he could dismiss ESP once he had met and liked J. B. Rhine. This time our letters crossed in the mail. In a letter dated August 12 I wrote him, "I'm sure you'll be pleased to hear that I don't want to be an astrologer, although I have no doubt that I could; instead, I want to write about astrology." And on August 14, he wrote me:

Yesterday, [I was] in a rather low state physically. . .. This is a background fact that intensifies my concern about the topic of my letter of yesterday.

I read the letter to Mother and she said she thought it fine except for one sentence she wanted me to leave out. "It will make Helen furious," she said, "and will do no good." The sentence expressed my confusion and concern that anyone as able, intelligent, and sensible as you are would devote any time, energy, interest (and perhaps even some money) to such fields as astrology and yoga.

Then in the afternoon I got your good letter of Aug. 12, and promptly tore up the other letter.

But not without quoting the offending sentence, perhaps verbatim! The opening shots had been fired, and for the next few years the debate continued via the U.S. mail. Neither one of us had the courage to discuss the subject of astrology face to face.

A year later, Dad sent me a copy of a little statement on "Sharing" that he had read at the Congregational Church in New Milford. In it he said:

Of all the laws of physical science, which is the most far-reaching? The one on the grandest scale? Surely the correct answer is the universal law of gravitation. Announced by Sir Isaac Newton about three hundred years ago, this law states that every particle of matter in the universe is related to every other particle in the universe, there being an attractive force between the two – between *every* two – a force which depends on the masses of the particles and their distance apart.

The significance of this law for us at this special service is that however far apart the two particles are, however small one or both may be, there is a measurable dependence of one on the other. This is a mutual – a truly shared – effect. Not only does the most distant star attract me, I attract the most distant star.

This struck me as a rather remarkable statement coming from my father, since one of his major objections to ESP was that it depended on "action at a distance."

I wrote him back:

I liked what you said about the interconnectedness of the universe. I wonder whether you noticed that your remarks are a beautiful argument in favor of astrology?

In 1970 it was really still the sixties. My interest in astrology was part of a resurgence of interest in Eastern religions, the occult, mysticism, and so on; and as a scientist, my father was both fascinated and concerned. He sent me an editorial from the *Saturday Review* that decried "a burgeoning cult of anti-reason" and the growing popularity of astrology, along with the I Ching, Tarot, numerology, palmistry, dream books, and Ouija boards, among other threats to rationality. I must have sent Dad something from Jung defending astrology, because he wrote back:

I was fascinated by the quotation from Jung. Too bad he didn't have a good math teacher who could make him understand the sense and power and grand generality of $a = b$, $b = c$, therefore $a = c$.[34]

The summer of 1970 my friend Marcia and I rented a cottage on Fire Island, and I fell in love with yet another totally inappropriate person. This one was a mere child: a deck hand on the Fire Island ferry. Nothing happened except a lot of unfulfilled longing, mostly for the adolescence I never had; but my bizarre passion did have one positive result: a sudden explosion of literary activity. There's a saying in the Weaver family: "All that force has to go somewhere." My frustration found its only outlet in my typewriter, which began clattering away for many hours a day. I began writing that summer, and I've never really stopped. That book – my version of *Lolita* – never got published and probably never will. You can't go to jail for what you're thinking; nevertheless, *Matthew* remains safely hidden in my files.

But that summer was the beginning of the rest of my life. I wasn't just tired of translating. I was tired of living in New York City.

Things happened: A guy punched me in the breast in the subway. I got mugged a block from my apartment. It got so I was afraid to walk to the corner deli for a quart of milk after dark. New York was getting scary. I began dreaming about the woods, wild animals, and sleeping on the ground.

My friend John Button said, "When civilizations are young, they value their cities. When they become decadent, they value nature." Was America becoming decadent?

Then, through Marcia, I met her poet friend Cynthia Poten, who described herself as "a little vegetable that had just arrived from New England." Cynthia took me camping, first in the Everglades and then in the Adirondacks. One perfect day in August 1971 we climbed Mount Marcy, the tallest peak in the east. On the way we paid our respects to the headwaters of the Hudson River, to which the Native Americans had given the poetic name Lake Tear of the Clouds. When we got to the top of the mountain we smoked a joint, and I had an epiphany. I looked down and I saw my path, and it led out of the city.

Some friends took me to Woodstock on a day trip. They let me out of the car on the corner of Mill Hill Road and Tinker Street, right across from the tiny village green. In front of me was a sign that read, in a funky New Age calligraphy: "Occult and Spiritual Books Upstairs: Also, Incense, Oils, Tarot Cards, I Ching Coins, Astrology Tools, and Miscellaneous Metaphysicals." I walked upstairs. It was like a little Weisers, the occult book store in the city.

Back in the city I had nobody to talk to about astrology except Marcia. Most of my intellectual friends put me down for taking it seriously. The poet John Hollander told me I was turning my back on my heritage. (Marcia, God love her, said, "He's turning his back on his.") Dick Howard was equally dismissive. Dan Wakefield and Allen Ginsberg were open to it, but nobody knew anything about it.

But on my very first day – nay, hour – in Woodstock I was suddenly surrounded by people who believed in astrology. I felt like I was in a foreign country, but one where I spoke the language.

Cynthia and I started looking for a house to rent. We subscribed to the *Woodstock Times*. Eventually we found a little house in West Shokan across the road from the Ashokan Reservoir, just a few miles from Woodstock.

The road we lived on was bordered by the tallest pine trees I had ever laid eyes on. I remember being overwhelmed by those trees. I was like a starving man in front of a big meal. I could only look at one tree at a time.

At first we went back and forth between the city and the country. Then I sublet my apartment.

On May 1, 1972, I moved to Woodstock.

That spring I sent my father a poem I wrote called "On the Bus to Woodstock:"

Hills you wait
the oldest
the wisest
I'm changing my home to you
for you I will be brave
so I can climb and fall and compass you

On the highway
slowly climbing
the air grows colder, and I feel your warmth
leap out of the frozen ground
gone, the city's poison clouds
the dun colors are becoming real
(woods of your foothills)
and rounding a turn I see, at the sky's edge,
that pale and gentle line
that means you are around me,
giving me the courage of my mind.

In my rush to return to you
I broke my mother's watch –
good, you'll keep the time
you'll swallow the sun like a pill
and send it back tomorrow through our pines

to pull me to my path.
You call the sun by name
and you know more stars than all my books.

I am of you
blankets of my soul
mother and father never to be mourned
well found in Woodstock
O hills at last I'm home

Dad wrote back:

You certainly have a gift for truly poetic expression. "The Bus to Woodstock" moves me very much. More on that topic later, the first chance we have to talk about a very tentative idea we have.

Of course, Dad being Dad, the tentative idea became, eventually, land and a house.

Meanwhile, now that I lived in the country I had to learn to drive – at the age of forty-one. Cynthia taught me in her Volkswagen camper bus, Mushroom. Dad bought me a white 1972 Volkswagen Super Beetle. I named her Sadie.

The whole time all these changes were happening in my life, I was keeping track of what was going on in the heavens. I was particularly struck by the connections between my chart and the charts of the people who were having such a profound impact on my life.

"Synastry" (from the Greek *syn*, together, and *aster*, star) is a fancy word for chart comparison, which the *Larousse Encyclopedia of Astrology* defines as "the comparative study of two or more charts of individuals, nations, corporations, and so on, for the purpose of determining compatibility, improving communication, or illuminating problems within a personal, political, or professional relationship."

The Moon's nodes are the two points where its orbit intersects the Earth's orbit (i.e., the ecliptic). Astrologers believe the Moon's nodes may give important clues to the life path of an individual for a given incarnation.

Most astrologers assume the validity of reincarnation. That dream I had when I was thirteen in which I visited a few of my past lives was one reason I was open to astrology in the first place. Looking back at that dream after I began studying astrology, I noted that it was a journey to the Moon that preceded those brief glimpses into those other lives.

For many modern astrologers, the sign on the south node of the Moon represents the qualities you come in with and the sign on the north node represents the qualities you need to develop in this lifetime. So node connections between charts are often found when two individuals are strongly connected in some way, or have something important to teach each other.

My north node – the qualities I came in to develop – is in Aries in the first house. Aries is cardinal fire, the sign the Sun enters at the Vernal Equinox, and the first house represents the self. Aries stands for what my father in his Scientist's Prayer called "the ever fresh and vital urge of spring." It stands for nature, and pure energy.

I remember noticing at the time that Matthew, the ferry boy whose innocent beauty bum's rushed me to the typewriter after decades of delay, had the Sun in Aries. So did Cynthia, who took me camping, taught me to drive, and got me out of the city. Each of these Aries people had something to teach me about unlocking my own energy and about living in nature instead of in my head.

I was particularly impressed by the mathematical precision of some of these connections between charts. Let's take Saturn contacts between charts as an example.

Saturn is the furthest body in our solar system that can be seen with the naked eye. For ancient astrologers, and up until the invention of the telescope, Saturn was thought to be the outermost planet, the big daddy of them all. So in astrology Saturn represents the limits of time and space. It is associated with time, old age, death, fathers, teachers, discipline, responsibility, order, structure, organization, government, and laws. In esoteric astrology Saturn is the Lord of Karma, the Cosmic Teacher. A very serious planet!

So Saturn contacts between charts are about deep lessons and the relationship tends to be permanent – or at least long-lasting – in its effects. It follows that Saturn-node contacts between charts are especially significant.

Marcia, the loyal friend who suggested I take classes with Zoltan Mason and who respected my interest in astrology when all the skeptics raised their collective eyebrow and thought I'd gone off the deep end, has her north node in Capricorn conjunct my Saturn within 27 minutes of arc. Her advice launched my study of astrology; and she often tells me that my talent for organization (my Saturn is in what Zoltan called "a good cosmic state") has always been an inspiration to her. The teaching goes both ways.

In the zodiac there are twelve signs of thirty degrees each, for a total of 360 degrees around the wheel. Each degree has sixty minutes, for a total of 21,600 minutes around the wheel. You don't have to be an expert in probability to see that 27 minutes out of a possible 21,600 is a very close connection.

Another example: Richard Howard, who got me my first translating jobs and taught me my craft, has his Saturn conjunct my Midheaven, the point on the chart that stands for career and profession. In this case the "orb" (degrees and minutes of separation) is not so close mathematically (one degree 38 minutes), but since the Midheaven (along with the Ascendant) is the fastest moving point on the chart, the conjunction "works."

But the most impressive of all these inter-chart connections is the one I have with Cynthia. Her first-house Saturn in Aries is conjunct my first-house north node within 28 minutes of arc. By helping me to reconnect with nature and making it possible for me to leave the city and live in the country, she literally changed my life.

If Saturn represents order, Uranus, the planet that overthrew Saturn's rule of the solar system when it was discovered in the eighteenth century, represents revolution. Astrologers find clues to the meanings of the modern planets in events at the time of their discovery. The eighteenth century was a time of upheaval. The invention of the telescope had changed science forever, and the industrial revolution and the American and French revolutions ushered in the modern age.

Uranus represents radical change in the life of an individual as well. Uranus takes approximately 84 years to complete its orbit of the zodiac, so some time in our early forties, Uranus will be opposite its position in our birth chart. Oppositions can be both uncomfortable and enlightening. Uranus has been dubbed "the habit-breaker," and this opposition can be a very disruptive energy in an individual's life. In fact, the Uranus opposition is the astrological signature of the so-called midlife crisis: the familiar phrase – and phenomenon – "life begins at forty."

At the time of my Uranus opposition, I left the city, learned to drive, and became an astrologer.

Helen in Woodstock 1973

Chapter Fourteen
WOODSTOCK

My view of life is that it's next to impossible to convince anybody of anything.[35]

Lewis Carroll

By the time I got to Woodstock I had been studying astrology for a little over three years. I wasn't exactly ready to put out my shingle; in fact, I have never felt comfortable charging money for astrological readings. Once money changes hands, people tend to take what you say more seriously, and that has always felt like a heavy responsibility to me. Saturn is very strong in my chart and I can't bear the idea of making a mistake: seeing something that isn't there, or not seeing something that is. We hippies have our standards!

From the very beginning I have preferred writing about astrology to giving readings. I managed to talk Geddy Sveikauskas, editor of the *Woodstock Times* and a confirmed skeptic, into letting me write a weekly column. "Correspondences" was not the kind of column most newspapers and magazines run, with mini-predictions for each Sun sign, as if there were only twelve possible biographies. Instead, I wrote a series of articles whose purpose was to teach astrology, to delineate interesting charts, or to give an astrologer's view of current events. The column became so popular (much to the disgust of the local astronomer, who also had a column in the paper) that Geddy actually began paying me for it. This, from a guy who was convinced his Moon was in Saugerties.

I began sending copies of some of my columns to my parents. In a letter to my father I said, "I hope you don't feel the nut has fallen too far from the tree." I still have the file folder where he kept our correspondence on this subject, marked "Helen and Astrology." In the fall of 1972 he wrote me:

Helen:

In one of your columns you say "Astrology believes that the universe is a meaningful system in which each element is part of a larger whole, and there is a connection between events in the heavens and events on earth." So far – so good.

However, when circumstances at a location A affect other circumstances at a location B, the viewpoint of science is that this influence is incredible and irrational – and in fact is not to be accepted as a normal cause and effect relationship – unless one can specify the *means* whereby circumstances at A are transmitted to and brought to bear upon circumstances at B.

The movement of the moon affects the tides, but one can calculate the gravitational forces that cause the tides, and the way these forces change when the position of the moon is altered, and these changes, in accordance with well established hydrodynamic theory, *explain* why the tides occur.

What transmits astrological influences? There are, in fact, relatively few recognized and analyzed types of forces – forces, that is, which can *act* at a distance: gravitational, electromagnetic, etc. *Why*, in *what* way, and acting *how* upon *what*, can the position of celestial bodies millions of miles distant, influence the temperament, the nervous constitution, or the potentiality, of human beings?

Love, Dad

Of course I couldn't answer this question. I still can't, and I'm not sure anyone can. I'm not sure "influence," with its assumption of causality, is the right word. I told my father that just because we don't yet understand *how* astrology works doesn't mean that it doesn't work. After all, do we really understand gravity? The normal growth of a cell?

Dad had started putting aside money in a special account to buy me land, and I had found what I thought was the perfect location. Riverby was conceived as a sort of anti-development development. A few miles from Woodstock, it features pristine woodland, with no lot smaller than five acres. The name was inspired by the Hudson River home of Catskill naturalist and writer John Burroughs. The idea was that a house on five acres could be heated with the wood from its fallen trees. There were all sorts of deed restrictions to guarantee privacy, the preservation of the land's natural beauty, and the protection of wildlife.

Those who purchased lots in Riverby were encouraged to camp out on the land before building. Tents and tipis were provided free of charge. My new friend, artist Michael Green, lived in a tipi year round, chopping wood and drawing water from one of Riverby's rushing streams. His lot was on the road he had christened "Middle Way" after the teachings of the Buddha. Riverby felt like the realization of my city dreams.

Dad wanted to know all about the price per acre and the approximate costs of clearing for a driveway, digging a well, running electric lines, taxes, and so on. I gathered all the information, but admitted that I was not yet able to contribute much in the way of money:

The transition which has been taking place in my professional life in the last couple of years from translator to writer and astrologer has eaten away at my savings, and I'm afraid my contribution to this project, at this stage in the game, will be a modest one. At the moment my funds are very low, due to the demise of Marc's shop [a little job I'd had at a record store] and the nominal nature of my earnings from the *Woodstock Times*.

I occasionally charge money ($10) for astrological chart readings that I do for people, but since I spend about 10 hours studying each chart, that is not as yet a very lucrative enterprise. I feel awkward about charging money for readings because 4-1/2 years of study and experience is really minimal for an astrologer. Even with my good Weaver brain and my passion for studying this is the hardest thing I've ever learned – much harder than learning the French language well enough to translate books. . ... I could make a good living off tourists if I were willing to lower my standards and put myself in the position of saying things I'm not absolutely sure of, but this of course I can't do.

I go on to say that I've decided to write a book about astrology – "an attempt to bridge the gap between the young people, most of whom accept it instinctively, and the skeptical older generation." I had even found a literary agent who was interested in the project and had offered me a contract.

Dad wrote back, thanking me for the information I had gathered, but worried about my prospects for earning a regular income:

I am (perhaps unnecessarily pessimistically) very skeptical of your having, within a reasonable time, any significant income from a book on astrology. I have written five books, and didn't get anything at all out of three of them. *Lady Luck* is the sole exception. . ...

Have any Riverby plots been sold in the last six months? Is the reputation of Woodstock (as a fairly "far out" community) a threat to the future value of land near Woodstock?

<u>Some</u> way we will work this out!

                                                        Love, Dad

I guess one of the reasons I am in a somewhat depressed state is that I have been seriously trying to read the astrology book you gave me – and the harder I try, the less I understand and the less I am impressed.

    The book in question was *The Case for Astrology* by John Anthony West and Jan Gerhard Toonder, published by Coward-McCann in 1970.[36] A few days later Dad sent me a letter he headed "Questions from an interested skeptic:"

Astrology, I assume, claims that the position of the planets at the moment of a child's birth has some effect upon, or is at least correlated with, the physical, mental, and emotional traits of that child.

If that configuration, at the moment of birth, actually <u>affects</u> the physical, mental, or emotional traits, how is that effect achieved? Is the inheritance of the child, the actual set of genes – the detailed molecular structure of the DNA which constitute its biological nature – different from what it would be if the "birth configuration" had been different? If so, what is the <u>cause</u> that achieves this effect? Can the cause be characterized in physical, chemical, or biological terms?

What <u>forces</u> are recognized by science – gravitational, electrodynamic, etc.? How do they depend upon the distance between the cause and the thing being acted on? Can there be an effect without a cause?

What sort of questions can astrology answer? Are there examples of questions whose astrological answers can be checked against the verified facts?

What position does astrology take with respect to the classical controversy between the relative importance of heredity and environment?

Consider three babies, born at the same moment of time, but one born in New York City, one in Madrid, one in Tokyo. Draw lines from each birth spot to the planet Mars and out into space: these three lines intersect the celestial sphere (how defined, incidentally?) at three very different places A, B, and C. Does this lead to three different astrological charts for the three babies?

Can you state, in one paragraph, what astrology claims that it can say about these three individuals – their native talents, their potentialities, their probable futures – or <u>what</u>?

If my comments, questions, etc., bore you or in any way irritate you, just say so and I will be silent. If you start to write anything about astrology, and want my candid comments or questions, let me be your confidential collaborator.

. . . The more I read in the West-Toonder "The Case for Astrology" the more confused I am:

"The planets exert no visible or detectable influence upon the events of the earth" (page 21)

"The real astrology is illogical but elegant" (page 22) etc. etc.

All my love to my mysterious daughter

Dad

My father's question about the three babies shows a kind of intuitive grasp of the geometry of the birth chart. Those three babies would indeed have three very different charts, because a chart is calculated not just for a certain time but also for a certain latitude and longitude. In all three charts the Sun, Moon, and planets would be in the same degree of longitude. But each chart would have a different Ascendant and Midheaven, to name only the two most important "cusps" on the houses, those slices of pie around the wheel. They were all excellent questions, but I didn't know enough to answer them yet.

Almost the same day I moved to Woodstock, I had met two gifted astrologers, Mary Orser and Allan Edmands, both of whom became good friends as well as colleagues. Together we founded the Woodstock Astrologers Guild, affectionately known as WAG. In April 1973 WAG attended an international astrology conference at NYU on the topic "Astrology and Science." In those days there was a lot of talk about putting an end to the "unnatural schism" between astrology and science that had developed in Western thought since the Renaissance.

The French psychologist and statistician Michel Gauquelin (1928-1991) spoke on something he called "planetary heredity." Something of a maverick – he was sometimes referred to as *"l'astrologue malgré lui"* (the astrologer in spite of himself) – Gauquelin devoted many years to painstaking research that showed significant correlations between the positions of planets at the birth of individuals and their choice of profession in later life. Although he was careful to distinguish his work from astrology, the distinction seemed more one of semantics than of substance. In my column, "Correspondences," I wrote:

Particularly impressive is his study of the position of the planet Mars in the birth charts of athletes. His results showed a frequency of Mars rising (on the Ascendant) or culminating (on the Midheaven) in the birth maps of athletes that significantly exceeded the level of chance distribution.

Gauquelin also found a correlation between the angular (rising or culminating) position of the Moon in the charts of writers, angular Jupiter in the charts of actors, and angular Saturn in the charts of scientists.

All of these results make sense to the astrologer. The nature of a planet rising or culminating does tend to color the individual's whole personality; and Mars, for instance, rules the musculature, physical strength and stamina, and goal-directed energy, and would naturally be emphasized in the horoscope of a professional athlete.

Interestingly enough, a group of Belgian astronomers, anxious to disprove the possibility of planetary influence, did a similar study of the planet Mars in the birth maps of athletes. Their findings confirmed Gauquelin's results so exactly that the scientists are now refusing to have the results published because they support astrology.

The story of suppression by the scientific establishment of facts and theories that run counter to its materialistic bias reads like a modern version of the Inquisition, and it is one that must soon come into the light of day.[37]

    I was on shakier ground when I reported on a talk by mathematician Theodor Landscheidt in which he offered a theoretical explanation of how planetary influence operates in terms of cybernetics, or information theory. This, of course, was one of my father's areas of expertise. I almost fell out of my chair when Dr. Landscheidt said that the existence of gravitational waves coming from the center of our galaxy into our solar system had been proved by Warren Weaver in *The Electromagnetic Field*.

    My father, to whom I sent this column, was not impressed:

Where did you get the idea that *The Electromagnetic Field* says anything about gravitational waves? It does not. Indeed I have to say that your article "Astrology and Science" leaves me very cold indeed.

You say "Dr. Gauquelin is a member of the French scientific community in good standing." I find no mention of him in any of my books. And who is Theodor Landscheidt? He is not listed in American Men of Science.

Perhaps you can explain all this to me orally. Come to see us! Love, Dad

In 1973 my father and I collaborated on a book entitled *Les trois étapes de la cosmologie*, by Jacques Merleau-Ponty and Bruno Morando, that I had agreed to translate for Knopf. I sent him my rough drafts of the text, rechristened *The Rebirth of Cosmology,* and he sent them back with his corrections and comments. To help fill in the considerable gaps in my knowledge of astronomy, he sent me a little classic called *The Nature of the Universe* by the British astronomer Fred Hoyle.

Helen at her Riverby Cabin 1970's

Chapter Fifteen
RIVERBY

It is not difficult to recognize a star if it is of the first magnitude.

Warren Weaver

In June 1973, with Dad's help, I purchased Lot 23 on Middle Way in Riverby: five acres of virgin forest at the meeting of two streams. Together we planned the construction, first of a tiny cabin inspired by the garden house he and Mother had built on their land in Connecticut, and eventually, of a house of my own.

My friend (and soon-to-be neighbor) Michael Green – a talented local artist who had turned his land into a kind of interfaith spiritual retreat – told me about the Golden Rectangle of the Greek mathematician Pythagoras, a formula used in the design of temples: the ratio of the short to the long side was the same as the ratio of the long side to the sum of the two sides. This formula was believed to produce the most harmonious and aesthetic result.

I asked my father to calculate the dimensions of the cabin based on this formula, and found a builder who was willing to work with some rather odd specs. We copied the design of the cabin, which was to have a little front porch facing a small clearing in the woods, from a vegetable stand with a shed roof on Route 28. The cabin was to have no electricity or running water and was to be heated by an Ashley wood stove. Some time after it was finished I wrote:

The Writer's Content in her Cabin

Here is ample time in little space;
silence enough to make me whole again;
a sweet simplicity that can erase
memory of past or fear of future pain.
Here are four white walls: north, south, east, west,
each with its window looking at snow and trees,
and in the corner of my hermit's nest

a cast-iron friend who will not let me freeze;
with bed, and books, and pots and hooks; a chair
for the rare visitor; a week's supply of wood,
paper and pens – oh yes, that carving there
is Michael's work. The light's especially good
in winter when the sun's below the eaves
and briefly, at noon, illuminates these leaves.

That sonnet was written after another unhappy love affair. But meanwhile, even as my father and I discussed the design of my future house, our debate on astrology continued.

Early in 1975 Dad presented me with a challenge. He managed to get the birth data – day, month, year, place, and even time of day – of Dr. Yuen Ren Chao, the first translator of *Alice in Wonderland* into Chinese. You may recall that in *Alice in Many Tongues* Dad describes Dr. Chao as "professor of oriental languages at the University of California, and very possibly the most perfectly bilingual Chinese-English scholar who has ever lived." He gave me his birth data to see what I could tell from his chart, as a sort of test of astrology.

I sent Dad my comments on Hoyle's book on astronomy and on Dr. Chao's chart in the same five-page letter. I began by quoting Hoyle:

"The most obvious question to ask about continuous creation is this: Where does the created material come from? It does not come from anywhere. Material simply appears – it is created. At one time the various atoms composing the material do not exist, and at a later time they do.

"This may seem a very strange idea and I agree that it is, but in science it does not matter how strange an idea may seem so long as it works – that is to say, so long as the idea can be expressed in a precise form and so long as its consequences are found to be in agreement with observation." (Astrology fulfills all these requirements!)

This is the very essence of faith. Yet at the end of the book Hoyle confesses his agnosticism and describes religion as "a desperate attempt to find an escape from the truly dreadful situation in which we find ourselves." This ending is in curious contrast to a book which describes an awe-inspiring universe characterized by continuous creation, constant energy, and a unity (for instance, the omnipresence of hydrogen) which makes it possible to construct valid theories by analogy. Yet when he faces the idea of personal death, this marvelous design becomes a "truly dreadful situation."

Hoyle goes on, "In [the Christians'] anxiety to avoid the notion that death is the complete end of our existence, they suggest what is to me an equally horrible alternative. [eternal life] . . . Now, what the Christians offer me is an eternity of frustration. And it is no good their trying to mitigate the situation by saying that sooner or later my limitations would be removed, because this could not be done without altering *me*."

So what's so bad about altering me? We think that if we have to give up our mind, that tells us "I'm so-and-so," we've given up everything about us. Christ and Buddha taught that we are something besides a mind which is obviously attached to a body, and I believe them. Who needs an eternity of being so-and-so? A lifetime is enough! But I feel one can *be*, without being so-and-so.

I feel that Fred Hoyle wants to keep on being Fred Hoyle after death, because somehow his view of the universe was a little too impersonal: his view of science was that he, Fred Hoyle, should keep his personal subjective feelings out of it. Well, I don't agree that that's a good idea, even if it were possible – and scientists are kidding themselves if they think it is. The fact that findings are relative to an observer should be acknowledged and included in the total picture. . .. All the truths we discover are meaningless unless they are relevant to us, unless they help us to live meaningful lives on this planet in the time we're here.

I've found astrology to be a language of symbols that helps me and other people to find meaning, and an overall pattern, in our lives. Not the comfortable feeling of everything being planned out in advance – I wouldn't find that comfortable, but some people might think they did – but a sense of an awesome and intricate order in which there is a force that operates, a kind of benign computer that is constantly sending out useful information to anyone willing to take the time to listen.

I've discovered that by assuming the existence of that benign computer and trying to remain open to its messages the quality of my life is improved, and I feel calmer in the face of almost any experience. This improvement and tranquility is enough to make me feel that I have a good working theory.

In my system, astrology is simply one of many ways or paths that may be used to enrich one's understanding of the few simple laws that govern the cosmos and to help one find one's place in it. It's one that appeals to a lot of people right now for many reasons. Science has made the mistake of disclaiming responsibility for questions of value, has paid its allegiance to the mind of man at the expense of his spirit. The result has been an abuse of machines and technology which has given science a bad name, and many older teachings that science has turned its back on are being re-examined.

But there are deeper reasons. Many churches lost their impact on people's lives by becoming too detached from the everyday realities of work and love. Religion didn't help a person to live with himself or with others. For a while it looked as if psychology was going to replace religion, but most orthodox analysts, in their anxiety for acceptance by the scientific community, made the mistake that science did and refused to deal with the human spirit.

I think Fred Hoyle is a deeply religious man who doesn't know it. I love his honesty and simplicity and I know I will read his little book again and again. I didn't know I was going to write a sermon!

Dr. Choo's [sic] chart is the chart of a serious scholar who must also be a charming man. Very independent, hard working, and ambitious. Natural language skills, a great facility for responding quickly to new situations, and a fantastic memory which is one of his major assets. Tremendous determination. A person who is concerned with communication and has an innate understanding of abstraction and symbols. Gregarious. A strong constitution, although I see the possibility of a health problem as a child that improves with age. There may have been poverty. Someone who dedicates himself wholeheartedly to his work, needs to contribute something useful to humanity, wants to shine at it, to be at the top of his particular field, and who has a very broad, humanitarian outlook. He feels most comfortable, most himself, when working at some occupation that serves other people. His kind of scholarship combines depth and thoroughness, the persistence of the true scholar who will track down out-of-the-way information, with versatility and breadth of vision.

I have concentrated on these psychological considerations in my study of astrology. If I were a different kind of astrologer, or one with more experience, I could probably tell you more about his background, his marriage, children, etc. Some of that's in the chart too but it's not my forte. I'd love to hear Dr. Chao's reactions some time.

Floor plan note: I want to include a small space at the top of my house for an "observatory" for a future telescope. Problem: heat rises and heat waves create visual distortion. My designer thinks maybe observatories have to be unheated but I suspect there's a solution. Do you know where we could look for information on this point?

Much love and please keep the astronomy coming – I'm <u>obsessed..</u> love, Helen

My father's reaction to all this was rather uncharacteristically ungracious. He headed his

response to my screed "Science – example of a scientific law:"

The gravitational attraction between any two masses m1 and m2 in the universe is a force, acting along the line between the two masses, proportional to the product of the masses and inversely proportional to the square of the distance between them.

No one understands why such a force acts. No scientist (certainly not Einstein) has ever explained this law.

But if you calculate out the force for the attraction of the sun on a golf ball, or of the moon on a drop of water in the ocean, and then measure this force experimentally, it comes out exactly right, each time. Reduce one of the masses by, say, 10.00%, and the force decreases by precisely that same percent, every time. The law is truly universal, it works exactly, every time, for a bit of butter attracting a speck of dust, or for a planet attracting a satellite.

Astrology: What requirements does it fulfill, with what precision and with what universality? And what and where is the evidence?

Telescopes: Do you realize completely that, as "looked at" through any telescope, from a tiny one a foot long, up to the 200-inch on Mt. Wilson, any star is simply a point (a pure, dimensionless, point of light), brighter for a large star and fainter for a small one, but only a point of light. A planet can, of course, be observed as a disk of light.

Putting a small telescope on the roof of a house has no advantage over having it supported on a tripod on the ground. You have to take a telescope up (by balloon or rocket) until it is above a large fraction of the atmosphere in order to gain any advantage from elevation.

Observatories are, in general, unheated to avoid the distortion caused by heat waves.

Your remarks concerning Dr. Chao seem to me to be just about what anyone would presume him to be from the well known facts about his life and activities.

Love, Dad

I didn't know anything about "his life and activities"! My reading was slapdash; I even misspelled Dr. Chao's name; but I don't think I did too badly for a beginner. But Dad more than made up for his rather dyspeptic reaction later in the year.

The Humanist magazine describes itself on its masthead as "a journal of humanist and ethical concern that attempts to serve as a bridge between theoretical philosophical discussions and the practical applications of humanism to ethical and social problems." In the fall of 1975 the editors decided to devote the better part of an issue to an attack on astrology. They prepared a statement condemning astrology as a "cult of unreason and irrationalism" and sent it to a selected list of distinguished members of the American Astronomical Society and the National Academy of Sciences for their signature. Among the 186 scientists who signed the statement were eighteen Nobel Prize winners, including Sir Francis Crick, Konrad Lorenz, and Linus Pauling, and among the non-Nobel laureates who signed it was our old friend Fred Hoyle.

Dad sent me a copy of the September/October issue featuring "Objections to Astrology" along with a little note saying that, as a member of the National Academy of Sciences, he had been asked to sign the statement, but had declined to do so out of respect for my feelings.

Chapter Sixteen
ASTROLOGY ON TRIAL

The scientific method requires of its practitioners high standards of personal honesty, open-mindedness, focused vision, and love of the truth.[38]

Warren Weaver

The statement in *The Humanist* that was signed by 186 "leading scientists" read as follows:

Scientists in a variety of fields have become concerned about the increased acceptance of astrology in many parts of the world. We, the undersigned – astronomers, astrophysicists, and scientists in other fields – wish to caution the public against the unquestioning acceptance of the predictions and advice given privately and publicly by astrologers. Those who wish to believe in astrology should realize that there is no scientific foundation for its tenets.

In ancient times people believed in the predictions and advice of astrologers because astrology was part and parcel of their magical world view. They looked upon celestial objects as abodes or omens of the Gods and, thus, intimately connected with events here on earth; they had no concept of the vast distances from the earth to the planets and stars. Now that these distances can and have been calculated, we can see how infinitesimally small are the gravitational and other effects produced by the distant planets and the far more distant stars. It is simply a mistake to imagine that the forces exerted by stars and planets at the moment of birth can in any way shape our futures. Neither is it true that the position of distant heavenly bodies make certain days or periods more favorable to particular kinds of action, or that the sign under which one was born determines one's compatibility or incompatibility with other people.

Why do people believe in astrology? In these uncertain times many long for the comfort of having guidance in making decisions. They would like to believe in a destiny predetermined by astral forces beyond their control. However, we must all face the world, and we must realize that our futures lie in ourselves, and not in the stars.

One would imagine, in this day of widespread enlightenment and education, that it would be unnecessary to debunk beliefs based on magic and superstition. Yet, acceptance of astrology pervades modern society. We are especially disturbed by the continued uncritical dissemination of astrological charts, forecasts, and horoscopes by the media and by otherwise reputable newspapers, magazines, and book publishers. This can only contribute to the growth of irrationalism and obscurantism. We believe that the time has come to challenge, directly and forcefully, the pretentious claims of astrological charlatans.

It should be apparent that those individuals who continue to have faith in astrology do so in spite of the fact that there is no verified scientific basis for their beliefs, and indeed that there is strong evidence to the contrary.

This statement was prepared by Bart J. Bok, emeritus professor of astronomy at the University of Arizona; Lawrence E. Jerome, a science writer; and Paul Kurtz, professor of philosophy at SUNY Buffalo, and editor of *The Humanist*. It's well written and sounds reasonable and convincing. It is, however, a mine of misinformation, and contains at least two outright lies.

The big lie is that there is no scientific foundation for the tenets of astrology. Scientists – particularly astronomers, who tend to be rather possessive about the heavenly bodies – are fond of saying this, but that doesn't make it true. What is true is that they are not *aware* of any scientific foundation for astrology because they have not taken the time to study the subject and indeed, do not consider it worthy of investigation.

There have actually been a number of tests of the astrological hypothesis: the assumption that there is a meaningful correlation between events in the heavens and events on Earth and, specifically, between the positions of the Sun, Moon, and planets at the birth of an individual and that individual's personality and character. Gauquelin's work is the most impressive in this field to date, but it is by no means the only work that supports astrology. But since no reputable scientific journal would publish results favorable to astrology, science continues to maintain that there are no such results.

To set up a meaningful test of astrology requires an understanding of the subject. If scientists are in charge, they lack this knowledge, and are seldom willing to acquire it; and if astrologers are in charge, the results are ignored: a Catch-22 that perpetuates the unfortunate schism between science and astrology that dates from the Renaissance.

The great astronomer Johannes Kepler (1571-1630), who discovered the three laws of planetary motion that bear his name, was also an astrologer. He wrote,

A most unfailing experience (as far as can be expected in nature) of the excitement of sublunary[39] natures by the conjunctions and aspects of the planets has instructed and compelled my unwilling belief.

The most persuasive argument in favor of astrology is experience. One need only study and work with astrology over a period of time to become convinced of its validity.

As my father so often reminded us, science is ultimately based on faith: "faith in the regularity of nature, in the inherent reasonableness of natural phenomena, in the discoverability of scientific laws." A hypothesis is nothing more than a belief that is under scrutiny and being subjected to experiment.

Granted, the astrological hypothesis is a curious one, and it is easy to see why a mind trained in the so-called "hard" sciences would find it difficult to entertain, especially given the vast distances between the Earth and the other bodies in question. As an astrologer, I am often awestruck before the mystery of the grand design of which astrology affords me a glimpse; but the fact remains that astrology works, even though to my knowledge no one so far really understands why.

The authors of the statement assume that astrologers believe that the "forces exerted by stars and planets" can "shape our futures;" that the "sign under which one was born determines one's compatibility or incompatibility with other people;" that there is a "destiny predetermined by astral forces beyond [our] control."

This seems to rule out free will altogether, but very few modern astrologers would agree that the planets represent "forces beyond our control." Astrologers believe that the positions of the Sun, Moon, and planets at the time of an individual's birth provide *clues* to that individual's personality and character, and like thinkers from Heraclitus to Freud, they believe that character is destiny; but these clues in no sense rule out free will. On the contrary, the insight astrology can provide, in the hands of a competent and experienced astrologer, can help individuals to realize their potential and live in harmony with their own nature – or, as the ancients might put it, to "master their stars."

If scientists tend to assume that astrology is a form of fatalism, this may be because the only contact they have with astrology is the predictions they see in newspapers and magazines. These little nuggets of advice for the day, week, or month, while they have some slight basis in astrological theory and practice, are necessarily based on Sun sign alone and hence have a very limited relevance to any one individual. Sun sign astrology is perpetuated for the simple reason that it is the easiest form of astrology to commercialize, since in this culture, at least, everyone knows what day of what month they were born.

But a glance at one's horoscope in the daily paper is no substitute for a reading with a competent professional astrologer based on a chart calculated for the exact time and place of birth, any more than a glance at an advice column in a magazine is a substitute for one or more sessions with a professional therapist.

The one sentence in the statement with which I agree is that "in these uncertain times, many long for the comfort of having guidance in making decisions." This guidance astrology can certainly provide; but not because it presents us with "a destiny predetermined by astral forces beyond [our] control." If that were true, there would be no need for decisions! No, the guidance provided by astrology *increases* our control over our lives by enriching our knowledge of ourselves and of the changing pattern of energies within which we live.

The little echo from Shakespeare ("The fault, dear Brutus, is not in our stars / but in ourselves that we are underlings") adds a touch of class to the statement. Of course, there's no telling what Shakespeare himself believed. He did put that dismissal of astrology into the mouth of Cassius, an unattractive character who was part of the conspiracy against Caesar. And the astrology Cassius dismisses is the fatalistic version that blames the stars for one's position in life.

One is tempted to believe that the bard put more of his own ideas into Hamlet, whose more nuanced belief system allowed him to remark, "There are more things in heaven and earth, Horatio, than are dreamt of in your philosophy." In any case, the list of great thinkers who did believe in astrology includes Pythagoras, Plato, Plotinus, Ptolemy, Aquinas, Kepler, Galileo, Goethe, and Jung. Not such shabby company!

In the absence of reasoned argument the authors of the statement resort to guilt by association: hence, the references to "magic" and "superstition" and the clear implication that most, if not all, astrologers are "charlatans."

The statement concludes by repeating that "there is no verified scientific basis for [belief in astrology], and indeed that there is strong evidence to the contrary." This is the biggest whopper of all, for no such evidence exists.

The key word here is "verified." For a hypothesis to be verified, it must be capable of replication. In fact, following the challenge laid down by *The Humanist*, a group of scientists calling themselves The Committee for the Scientific Investigation of Claims of the Paranormal (CSICOP) did embark on a replication of Gauquelin's "Mars effect" for the sole purpose of debunking it (not exactly the proper attitude for open-minded research).

And when, to their dismay, their results supported Gauquelin's findings, they tried to cook the statistics and fearing exposure, ultimately refused to publish their results. The whole sad story of this shameful coverup, which was published by astronomer and whistle blower Dennis Rawlins in *Fate* magazine in October, 1981, can be found in the updated edition of John Anthony West's *The Case for Astrology* and in Suitbert Ertel and Kenneth Irving's, *The Tenacious Mars Effect.*

In the 1950s an American psychologist named Vernon Clark (1911-1967), intrigued by recent work in biology that lent support to the astrological hypothesis of a celestial-terrestrial connection, took the rare step of deciding to study the subject for himself. He began casting horoscopes and, like most people who take the time to do this, he began to suspect that there was something to it.

He decided to put astrology – or rather, astrologers – to the test. Because of the complexity of astrological analysis and the number of interlocking variables in the birth chart, he took a holistic approach. Instead of isolating a single variable, as Gauquelin was doing in France, he decided to test the ability of astrologers to distinguish between individuals on the basis of the birth chart alone. He devised a series of blind trials which, despite a few minor flaws, provided an early model for further experimentation.

In the first test, twenty astrologers were asked to match ten birth charts of real people, five men and five women, with ten mini-biographies describing career, marriage, hobbies, and health. To eliminate possible clues he chose subjects who were all born in the United States (e.g., a chef was eliminated because he was born in France). He made sure that each subject had been well established in his or her field and had a reliable time of birth.

In the second test, twenty astrologers were given ten pairs of charts and ten case histories and asked to pick the chart that fitted the case history. They were not told that one of the charts was cast for a real person and the other was a spurious chart cast for a random time and place.

In the third test, thirty astrologers had to distinguish between each of ten pairs of charts, one of a person of high intelligence and the other of a victim of cerebral palsy.

All three tests were given to a control group of psychologists and social workers who had no knowledge of astrology. In all three tests, the astrologers performed at a high level of significance, a few achieving perfect scores, and all performing above chance. The control groups performed exactly at chance. The results of the first two tests, with slight variations, were replicated in the 1970s by both Dr. Zipporah Dobyns and Joseph Ernest Vidmar, the latter with results at a high level of significance.[40]

I had been aware of the Vernon Clark tests since October, 1970, when the first of the three, the matching test, was reproduced in the astrological magazine *Aquarian Agent*. I had always planned to try taking the test myself but wanted to wait until I was a better astrologer.

In the fall of 1976, smarting from the sudden ending of my latest ill-advised love affair, I limped off to Connecticut and holed up with my parents for a while. I had started interviewing my father on tape in the spring of that year, and we resumed work on that project. And feeling that I was now as ready as I'd ever be, I began studying the ten birth charts of the Vernon Clark matching test, and comparing them with the ten biographies.

As I sat at my father's desk with the charts in front of me Dad started looking over my shoulder. I asked him if he thought this would be a valid test of astrology, and he said he thought it would. I asked him to calculate the probability of getting a score of 100% on the matching test on a purely random basis. I still have his figures, showing the probability of getting all ten right to be one in 17,625. I told him that astrology was not an exact science. Like probability, it had statistical validity. I could see that he was taking that in.

He was fascinated by the idea of the test. After all these years, I finally had his attention, so I knew I had to do well. I studied those charts for weeks, taking notes on each one as if I were going to give ten different chart readings.

I remember one chart in particular that had the Moon in Gemini conjunct Saturn in the sixth house. Right away I suspected that that must be the chart of the man who suffered from asthma: The Moon rules childhood, asthma is a childhood disease, Gemini rules the lungs, Saturn means limitation, and the sixth house is all about health. I myself, with three planets in Gemini and an "afflicted" sixth house, had had asthma as a child. These were the kinds of clues that helped me match up the charts.

At last I was done, but was frustrated by the fact that I didn't have access to the answers. I knew that they had been reprinted in the November, 1970 issue of the *Aquarian Agent.* I wrote every astrologer I knew and eventually, thanks to the generosity of Al Morrison, a complete file of the Vernon Clark material, including the answers, arrived in the mail.

I had eight out of ten right. I would have had a perfect score, but had switched two of the women at the last minute.

I told my father, "The women were harder than the men."

"Why is that?"

"Because the birth chart shows potential. These women were all born in the early twentieth century, and it was much harder for women of that generation to realize their potential."

My father, who had a file labeled "Women in Science," was aware of the dearth of opportunities for women at the top of any profession, and deplored it.

He said, "Aha!"

Warren, Helen and Pluto circa 1976

Chapter Seventeen
THE DIMENSIONS OF PERSONALITY

Science at a brash state of arrogant youth often is contemptuous of ancient wisdom. But time after time science has had to agree that the old truth, although often muddled in form and perhaps falsely substantiated, was in sober fact a valid truth. [41]

Warren Weaver

My father was impressed that I had done so well on the Vernon Clark test, but he was not that surprised. As a matter of fact, he had become intrigued by the astrological hypothesis *before* I got the answers to the test.

I had told him about the work of Frank Brown, a professor of biology at North Western University in Evanston, Illinois, who found a strong connection between the life cycles of various organisms and the phases of the moon.[42] And I'm sure I had mentioned the work of John Nelson, a meteorologist with RCA who had found a significant correlation between the kind of storms that interfere with radio broadcasts and the heliocentric aspects of the planets, especially the so-called "hard" aspects of astrology.[43] Neither of these men had any interest in astrology, but their findings indirectly supported the astrological hypothesis of a connection between events in the heavens and life on Earth.

Dad had found the original results of the three Vernon Clark tests persuasive; but I think the key to his gradual opening to the possible validity of astrology was his respect for my intelligence and the seriousness of my attack on the matching test: the sheer number of hours I put in studying those ten charts.

While I was working on the test and afterwards, while I was waiting for the answers to arrive, we continued to record our conversations on tape, but they were no longer limited to Dad's stories about the objects he had collected on his travels. We began discussing the relationship between astrology and science and we got into some pretty heady territory: quantum physics, relativity, and the nature of the universe.

Unfortunately, some of our most interesting conversations did not get recorded. But I do remember Dad looking over my shoulder as I was studying one of the charts and asking me, "What's the geometry of that?"

I explained that the birth chart was a two-dimensional map of a three-dimensional display: the state of the heavens at the moment of birth.

"What are those angular slices of pie?"

"Those are the houses."

"Are they in the sky?"

"No, they have to do with categories of experience on Earth. They have no geometric location. They are a fairly abstract notion, and yet where you locate them is important."

"What's that thing that looks like a brandy snifter?"

"That's Pluto."

I taught him the celestial geometry of the birth chart, using the beautiful old library globe that had stood in his study for as long as I could remember. He taught me physics; I taught him astrology. He was the better student!

At a certain point he asked me to show him an elementary astrological text. I chose Margaret Hone's *Modern Text Book of Astrology*, an old British warhorse published in London in 1951. The book was dedicated:

To Charles Carter, President Emeritus of the Astrological Lodge of London, The Headquarters of Astrology in Great Britain. Through his careful and scholarly writings, the "Royal Art" has flourished in this century, has been purged of many of its superstitions, and has gained the respect of all who give time to its study.

It was a good choice. Dad took the book into the living room and sat in the old brown chair next to Mom's couch and across from the TV, the chair he used to sit in every evening with his detective story and his Old Fashioned.

I can still see him sitting there, reading Hone's description of the predominately

Cancerian person (his birthday was July 17):

Towards people he likes, he will be sympathetic, protective and guarding, while as regards things, he will have a natural instinct towards collecting. This may lead to the collection of any rare and interesting things, or may be no more than the urge to let things collect, so that clutter never gets thrown away. He is a great home-lover. Like the crab, he needs the shelter of his own shell. He is patriotic about his own country, and devoted to his own family.

As Dad read this description of his sun sign I heard him mutter, "Well, for Pete's sake!. ... Well, for the love of Pete...." And I saw him shake his head, as if in disbelief. In this book of astrology, of all things, he was recognizing himself.

And I was watching the last traces of his lifelong resistance fall away. I think it may have been that bit about collecting that got him. His Lewis Carroll collection was, after all, the second largest in the world. But my father also collected interesting words, including terms of venery; knives; wine; oriental art; etchings; seashells; autographs of scientists; and the doodles that were made during the endless meetings he had attended, some by famous people and heads of state.

That "for Pete's sake!" of Dad's was probably the single most amazing moment in my life. It felt like a miracle. What were the odds? And our ongoing dialogue felt miraculous too. For the first time in our lives, my father was asking me serious questions about astrology, not in order to trip me up, but because he really wanted to know the answers. And for the first time in our lives, I knew enough about the subject to be able to give him intelligent answers. We were both excited by our dialogue, and we felt that it was important: A scientist and an astrologer were actually sitting down together and talking, with mutual curiosity and respect.

It was Dad's idea that we write a book. He wanted to call it *The Dimensions of Personality*. He started making lists of personality parameters: extravert/introvert, selfish/unselfish, stubborn/pliable, emotional/logical, adventuresome/cautious, moody/gay, and so on.

Interestingly, this was very much like what Michel and Françoise Gauquelin were doing in an extension of his research on the Mars effect. After first focusing on the correlation between *professions* and planets – sports champions and Mars, actors and Jupiter, scientists and Saturn, writers and the Moon – the Gauquelins went on to study the possible link between the so-called "Martian," "Jovial," "Saturnine," and "Lunar" *temperaments* and the corresponding planets, with results that replicated their earlier findings.[44]

Dad thought that at least part of our book should be in dialogue form. Dialogue, after all, was a traditional form of exposition in the history of Western philosophy, from Plato to Galileo. So I'd like to honor his wishes and use that format when appropriate (and when I have that conversation on tape).

For example, when I began work on the Vernon Clark test, he wanted to know:

WW: Can you tell by looking at a chart whether the person is male or female?

HW: For most astrologers in the West, that would be very difficult. When Vernon Clark devised his experiment, he included sex with the data he provided to the astrologers taking the test.

I understand that Hindu astrologers have an accurate method of predicting the sex of a child from the chart of conception, but I haven't worked with conception charts myself, so I can't report on that.

You *can* get from a chart a very good impression of the traditional secondary sex characteristics: e.g., how aggressive or passive a person is likely to be. If a woman has what used to be called "masculine" traits, that will show up in her chart, and if a man has the so-called "female sensibility," that will show up too. Homosexual or bisexual tendencies are relatively easy to recognize. But the actual physical sex is not something Western astrologers can arrive at with any accuracy, as far as I know.

WW: Since the genetic material is laid down when the sperm fertilizes the ovum, why don't astrologers use time of conception rather than time of birth in casting a horoscope?

HW: That's a very good question, and one that astrologers themselves have been asking for years. I suspect that the main reason Western astrologers work with birth time rather than conception time is that there's no reliable way of determining time of conception. In most cases, the information is simply not available. Even if a mother were quite sure she knew the date on which conception occurred, the actual moment of fertilization would remain unknown.

On the other hand, there's nothing arbitrary about the moment of birth. After conception and death, birth is certainly the third most critical moment in the life cycle of a human being. It is a dramatic event which marks the beginning of that individual's existence as an independent organism, and perhaps the single moment in their life when they are most helpless, most impressionable, and therefore most receptive to the influences of their new environment. It is the biological, social, and legal starting point of an individual's life on this planet.

Finally, the best argument in favor of birth charts is that they work. Planetary configurations at the moment of birth do provide meaningful clues to temperament and character.

There is some evidence for the belief that conception charts – and even death charts – are also significantly related to personality. I understand that conception charts are widely used in Eastern astrology. Hindu astrologers cast two charts, one for birth and one for conception, arriving at the latter by a somewhat arbitrary method which need not concern us here. I have heard that with these conception charts Hindu astrologers can predict events in an individual's life with a high degree of accuracy.

WW: What is the zodiac?

HW: The word comes from the same Greek root as "zoo" and "zoology;" literally, it means a "dial of animals." The signs of the zodiac were inspired by constellations, many of which the ancients identified with animal symbols: Aries, the Ram; Taurus, the Bull; Cancer, the Crab, and so on.[45]

The zodiac is an imaginary elliptical band that corresponds roughly to the ecliptic, the Sun's apparent path around the Earth. Modern astronomers put it at nine degrees on either side of the ecliptic. It is within this relatively narrow zone of the heavens that all of the planets can be seen from Earth as they follow their orbits around the Sun. This elliptical band has been divided into twelve segments of thirty degrees each, and each of these thirty-degree segments bears the name of one of the twelve signs of the zodiac.

Dad saw our book as doing for astrology what *Lady Luck* did for probability: that is, it would be an introduction to the science that was understandable by the layman. I pointed out that in the case of *Lady Luck* he was dealing with a relatively new science, whereas astrology was a very old science that had fallen on evil days. Dad thought there should be a chapter on the history of astrology. He said, "Anything that has had a long history can't be dismissed lightly."

He thought the book should also include an introduction to the symbolic language of astrology, a basic explanation of its elements. What information do you need to make a chart? How is this information expressed geometrically? What information does the chart provide?

We kept coming back to his original question, the great stumbling block: How can planetary influence operate over such vast distances? Dad said, "It is to scientists essentially incredible that a small difference in the angular orientation of exceedingly small forces could have an effect on the organization of the genetic material, or on the way in which the amino acids make up the DNA of an individual."

We returned to this mystery again and again, and while I cannot say that we solved it, thanks to my father's training in physics, we did arrive at a resolution of sorts.

For the time being, we agreed on an idea that seemed to bridge the gap between astrology and science: the idea that nature is all one piece of cloth, that everything is connected, and that any change anywhere in the whole affects everything else. At lunch that day Dad told Mother that the basic tenet of astrology was the fundamental unity of the universe.

He said that there were only four kinds of forces recognized by modern physics: strong forces that operate on a molecular level; electromagnetic forces (including radio waves); weak forces; and gravitational forces. He said that these four were the only known ways that the state of matter could affect the state of other matter. He felt that none of these could play a role in astrology.

"What is the difference between energy and force?"

"A force is an entity that causes the acceleration of a mass. Force equals mass times acceleration. Energy is the capacity for doing work. Work equals force times distance. Energy is a broader idea than force."

Dad felt that electromagnetic forces were too small and too evanescent to serve as the actor in a cause and effect relationship between planets and individuals.

"The effect of the moon on the tides is gravitational. The effect of the moon on organisms that spawn according to a cyclical pattern [as in the work of Frank Brown] is a mystery. The mechanism is not understood."

Dad felt that the "force" operating in astrology might be one of these four, or a fifth force as yet unknown, or not a force at all. He thought the mechanism could be something much broader and perhaps not characterized as a force. But it would be the motivating factor in a cause and effect relationship.

He said, "The cause and effect relationship involved in the process of growth, say on the part of a one-celled organism, is not understood. The principle of growth remains, essentially, a mystery.

"The more you learn, the more mysteries you encounter. Humility is called for."

He said that he was fascinated by these ideas: more interested than he thought I could possibly realize. He said that he worked on our book in his sleep, or when he woke up in the middle of the night.

Warren and a model of DNA 1953

Chapter Eighteen
WARREN WEAVER ON MODERN PHYSICS

I find the present situation of physics with respect to so-called elementary particles essentially intolerable. I do not believe that the physical world need be described in a way that leads to such continuously ramifying complexity.[46]

Warren Weaver

Here's a transcript of a typical conversation that did get on tape. It started by my reminding my father of his old question about scientific law: "If a miracle happened, would you think it represented a temporary suspension of one or more of the laws of nature, or the existence of a law as yet unknown?"

WW: Traditional science is very fond of talking about cause and effect, and indeed is very dubious about discussing any effect unless it is possible to correlate it with some identifiable cause. And that cause, in a great many instances – not all, but in a very large number of instances – operates via some force. That doesn't explain the cause, to say that it produces a force; but the force is something that can be characterized precisely and its consequences worked out; and then you can find out if the consequences do check with the experimental facts.

This traditional procedure of classical science has been extremely successful, for example, in the whole field of dealing with electrical phenomena. On the other hand, it's interesting to see the point that this procedure has reached in modern solid state physics, in trying to describe what matter is composed of, and how it is composed.

This has gone so far that we've now gotten to where we have a whole set of families of so-called elementary particles. (Incidentally, it should be noted that this is practically speaking a contradiction in terms, because if they are elementary particles, there shouldn't be a whole lot of different kinds of them.)

It was all very neat when we thought that all matter was composed of molecules, and the molecules were composed of atoms, and the atoms were composed of a nucleus with attendant electrons moving around it. That was a very neat and proper kind of description – assuming, of course, its validity and its efficiency. That's the kind of statement that science *ought* to have as a basic theory of the construction of matter.

But what's happened to that? Well, that's just kept getting more and more complicated. We've got more and more phenomena that weren't handled by that kind of a picture, until now we have not two or three kinds of elementary particles, but we've got whole sets of families of elementary particles. And the ridiculous character of it is indicated if you just simply recite the names of these things. It may work, but it's just obviously silly: hadrons and leptons and snarks and antisnarks and quarks and antiquarks. And some of these particles are assigned properties that are so strange that in point of fact, one of them is called the particle's "strangeness." That term is actually used. We speak of the "strangeness" of a certain particle. We also speak of the "color" of a certain particle or the "charm" of a certain particle.

HW: And these are technical terms?

WW: These are technical terms.

HW: It sounds like they've gone round the bend.

WW: Well, it's really a *reductio ad absurdum* except for one thing: That with all of this complication, it seems to enable you to explain the way that matter is constructed and how it operates, in the sense that, if you say how the experiment is going to be conducted you can tell how the experiment's going to come out, and it does.

In other words, except for empirical success, this has just reached nonsense, it's just *Alice in Wonderland* all over again.

HW: And of course one of your balmy mathematicians thought up that idea.

WW: That's where modern particle physics has gotten, so if it's a question of being peculiar and using strange words and involving essentially unimaginable things, science can't throw anything out of the window as being non-scientific for those reasons.

HW: That makes me feel a little better about the present state of astrology.

WW: Of course, I've always felt this way about the whole quantum theory. I didn't believe any of it. I thought it all sounded crazy, and that it was going to be highly regrettable if it turned out that it worked. Max [Mason] and I used to be absolutely convinced of this. We wouldn't even use the language of quantum theory except in scorn. Neither he nor I really believed any of it.[47]

HW: How do you feel about it now?

WW: I don't believe it any more than I did then.

HW: Does it work?

WW: It works, just embarrassingly well. There are just millions of experiments that can be described in advance and the results predicted, and when they're performed, the results come out that way. There's no question about that. I concede that; but I don't concede that I like it, or that I think it's a theory that has any artistic merit whatsoever. I can't believe that the world is built that way.

HW: Somewhere you talk about a scientific law as one that has universality, that enables you to predict with great precision, and your predictions will come true every time.

WW: Not every time. There are lots of scientific laws whose results are statistical in character. They say if you do something or other, then such-and-such a percentage of the time it will come out this way; but not necessarily every time.

There are two ideas about light that are held simultaneously by all physicists nowadays. Although the two contradict each other, every physicist uses both of them. One is that light is composed of particles, and the other is that light is composed of waves.

HW: Well, if they have two different terminologies and they're contradictory, then what do they do–use the one that works the best in a given situation?

WW: Yes, they use the one that's handiest in a given situation. The wave theory is one for which there's an awfully good mathematical technique, and by wave theory you can work out what's going to happen very neatly and very powerfully. The particle theory is harder to deal with, and so more often than not physicists use the wave theory of light rather than the particle theory of light. But they use whichever one fits in best with their ideas of a given experiment.

But the two are inconsistent, and every honest physicist knows that they're inconsistent, that it's just unthinkable that they would both be true. It's just unthinkable that light would be both one and the other simultaneously, but nevertheless this inherent mix up in language and complication in language and ridiculous contradiction in language is entertained by quantum theory people. They just pay no attention to it.

HW: Is this one of the reasons you don't like it? Because it contains this inner inconsistency?

WW: That's right. Hasbrouck Van Vleck [1899-1980] is one of the world's experts in this and a great physicist, and a complete believer in quantum theory, and he thinks that I am just unspeakable because of what I say about quantum theory. But I expect quantum theory to be completely overhauled – perhaps to completely collapse under the weight of its own inner contradictions – until the time will come when some really great physicist will say, Well, enough of this, now, come on. We've swallowed this stuff long enough, let's start all over, and let's try to get a decent, coherent, artistic way of describing nature.

Max and I worked on that. We had ideas about a basic unitary law of nature that would completely replace quantum theory. The trouble was, we just never could get anywhere with them.

One of these ideas was something like this: that every particle in the universe sends out a message to every other particle in the universe at certain intervals. And these messages are of two kinds. One is, so to speak, a request, and the other is a command.

As a matter of fact, we said that the law of gravity is nothing in the world but the total recognition of every particle in the universe that it's not alone: that this universe is filled with stuff, and that when a given particle says something, it goes everywhere, and everything else hears it, and talks back to it; and its recognition of the fact that it is not alone in the universe is nothing in the world but the general law of gravitation.

And the effect of this on a particle, the effect of having this consciousness, so to speak – consciousness being an impossible word in this connection, of course, because this is just a fine single particle of matter that I'm talking about – this consciousness that it is not alone in the world: the particle responds to that by being attracted to everything else in the universe, according to a law that varies in a certain way. But the basic philosophical reason for that force, Max and I said, is simply this recognition on the part of the particle that it's not alone. If there were only one particle in the universe, there would be no gravitation at all.

Well, we worked and worked and worked at trying to get ideas along this line that would make some kind of sense, and we never could get anywhere with it.

HW: Could you make an analogy between those two kinds of messages, the requests and the commands, and a positive and negative electrical charge? Or don't you see it that way?

WW: Not quite, but that isn't too far away from fragments of ideas that we had.

HW: You've been talking about these particles sending messages, these little tiny pieces of matter sending messages, and at one point you mentioned the word "consciousness." And you said, But of course we can't talk about consciousness in connection with such a tiny little piece of matter, and I thought of two things.

There is a theory – I don't know if it's espoused by any scientists or not, but there is a theory that I've heard that *everything* has consciousness on some level: that there's some sort of a vibration going on. Obviously, a rock is going to vibrate at a much lower rate than an amoeba. At any rate, this theory exists. And I wondered if the idea that it is possible for an elementary particle to have consciousness has, in fact, been scientifically disproved, or if it's just simply one of those ideas that's been classified as unthinkable, by definition.

WW: I think the latter. It hasn't been disproved. In modern particle physics, the way in which the forces operate is totally different from our ordinary idea of a force.

Dad explained that when two particles interact something is exchanged which is not energy, but an actual piece of matter; and these effects take place in ten to the minus 23rd power seconds.

WW: To talk about things happening in ten to the minus 23rd power seconds is just a misuse of language, it seems to me.

HW: Because there's no way that we can measure those events with our sensory equipment? Because the scale is so completely foreign to our experience?

WW: The scale is utterly foreign to our personal experience, or anything that ever could be our personal experience.

HW: So it's like fairyland again.

WW: Again. You can talk about it, but it's nuts.

HW: Is this elementary particle theory part of quantum physics?

WW: Yes.

HW: So this is more madness, as far as you're concerned.

WW: Utter madness, except for the terrible fact that it works.

Chapter Nineteen
THE UNITY OF THE UNIVERSE

A very great deal more truth can become known than can be proven.[48]

Richard Feynman

Getting back to astrology, Dad wanted to know whether the circle at the center of the birth chart represented the Sun.

HW: That circle symbolically represents the Earth, or more precisely, the center of the Earth.

Scientists, especially astronomers, who dismiss astrology as pseudo-science, tend to make much of the fact that astrology is Earth-centered. They see astrologers as naïvely clinging to an outmoded conception of the universe that predates the discoveries of modern science, especially the heliocentric solar system of Copernicus.

It is true that astrology developed within the context of a geocentric conception of the universe, and even today a great many (though not all) astrologers work with an Earth-centered diagram of the heavens. They are interested in studying the manner in which the heavens relate to an individual on Earth, and to this end they use the apparent motions of the Sun, Moon, and planets in relation to the Earth as timing devices.

The great astronomers who first worked with the heliocentric solar system – Tycho Brahe, Johannes Kepler, and Galileo – were all practicing astrologers. And because they understood astrology, they knew that the new world view in no way conflicted with or invalidated its theory or practice. Copernicus himself, although his precise opinion of astrology is unknown, nowhere in his voluminous writings even implied that the heliocentric system had overthrown astrology.

I read Dad a passage from a little booklet called *The Basic Principles of Astrology* put out by the American Federation of Astrologers:

HW: "In its essential nature astrology is not dependent upon *any* system of the universe, but has relied rather on the actual observation of the planetary bodies as they appear to move from a point of reference at any given position on the earth. This may be called *the principle of apparency.*

"It is as old as astrology itself and as new as Einstein's theory of relativity, with its assumption of a point of reference which may shift according to the convenience and the intention of the observer, and from which one may observe the apparent movements of other bodies."[49]

WW: So the correlations that astrology uses are based upon the *apparent* motions of the heavenly bodies, and therefore any confidence in those correlations isn't shaken by the fact that somebody comes along and says, Well, those weren't the real motions; the real motions were different. Well, all right, so they were; but I never was depending upon the real motions. I was depending only upon the apparent motions, that's what I based this all on. The apparent motions worked as timing devices.

HW: Yes.

I read Dad another quote from the booklet, this one bearing on our ongoing problem of action at a distance.

HW: "In an age when one reasons from cause to effect, [astrologers] may speak of 'the influence of the stars.' Due to the vast distances involved, this concept was held to be scientifically ridiculous for a long period of time. However, as John J. O'Neill, the late science editor of the New York Herald Tribune, wrote:
'The hypothesis of the astrologers that forces are transmitted to the earth without attenuation with increasing distance, and do not vary with respect to the differences in masses of the sun, moon and planets on which they originate, was totally inconsistent with the old style Newtonian mechanics; but today it is in complete accord with the much more recent Einstein photoelectric theory, which demonstrates that *the effect of a photon does not diminish with distance* [emphasis mine], and which has been universally adopted by scientists to supplant the Newtonian mechanics in that field.'"[50]

Would any kind of electromagnetic waves anywhere along the spectrum weaken by the time they got to Earth – including waves of visible light?

WW: No. That's just the point. You see, a photon, which is a unit of light, starts out being a photon, and no matter how far it goes, it's always just a photon.

HW: In other words, it's like a calorie: it's a unit of measurement?

WW: That's right. And it's a unit of measurement of something which is essentially indivisible.

HW: Are we talking about visible light, or ultraviolet, and so on?

WW: Either. A unit of light, no matter what its wave length is, is called a photon. And you see, back of this is this damnably confusing and almost intolerable position of quantum theory that light is both a particle and a wave. The photon is the particle aspect of it, and that's just a little bullet of light, and that bullet of light, no matter how far it goes, is still just one bullet of light.

But if you try to put this in terms of wave theory, then you're in a dilemma, because we don't know of any waves that do not attenuate as they pass through any medium. The waves get weaker.

So the dilemma is a dilemma which is inherent in modern quantum theory. I ordered today these lectures on physics by Dick Feynman. I want to know whether he has managed any kind of resolution of this conflict between wave theory and particle theory.

You see, one of the amazing facts is that these two ways of looking at phenomena turn out to be mathematically identical. You have exactly the same mathematical equations that you write down and manipulate in one case that you do in the other.

HW: In that case it doesn't seem as if the inconsistency is so serious.

WW: And they always lead to the same result.

HW: I would think that you would be able to mathematically prove that there was no inconsistency. It sounds like two sets of language, if they're both translatable into the same formulas.

WW: Incidentally, Einstein's photoelectric theory was his first great triumph, long before relativity. He was known to physicists and highly respected by them for that long before he got to relativity theory.

Well, I will certainly concede that the modern scientist has to admit that because of that basic fact that the effect of a photon does not diminish with distance, you've got physical phenomena going on in the universe that do *not* attenuate with distance.

HW: We've got at least one example.

WW: Yes. No scientist has a right to claim that everything has got to attenuate with distance, because there's a very significant example in which physicists now agree that that isn't so. And therefore to say that I just can't conceive of anything happening out at the stars that would still have enough oomph left in it by the time it gets all the way down here when a baby is born to have any effect on the personality of that creature. If a physicist says that, you'll just have to remind him that we have at least one significant example of such a phenomenon.

HW: In an article in *The Humanist* entitled "A Critical Look at Astrology," the astronomer Bart Bok writes,

"The known forces that the planets exert on a child at the time of birth are unbelievably small. . .. The stars are so far away from the sun and earth that their gravitational, magnetic, and other effects are negligible."[51]

WW: Well, of course, this is because he's thinking only of influences that are comparable to gravitation, which does attenuate.

HW: He rules out magnetic and radiative forces as well. He says that the sun's radiation is "far in excess of the radiation received from the moon and all the planets together." Yet Nelson's work shows that the sun may be a sort of mediator of the planets' effects, in a complex interrelationship.

Bok goes on,

"It seems inconceivable that Mars and the moon could produce mysterious waves or vibrations that could affect our personalities in completely different ways. It does not make sense to suppose that the various planets and the moon, all with rather similar physical properties, could manage to affect human affairs in totally dissimilar fashions."

But the different planets have different mass, different distances from the Sun and from the Earth, different speeds of rotation and orbit, different chemical composition, different temperature, different atmospheric conditions.

WW: Yes, and really underlying everything you're saying, which I think is quite justifiable, is the idea – I haven't yet written it down in good form, but I'm going to, at least in as good form as I'm capable of. One night, I had a very good formulation of this in the middle of the night. I had Assumption A and Assumption B.

HW: Can you remember it?

WW: No, I can't; I couldn't even get them the next morning; but the essence of Assumption A is an affirmation of belief in the essential unity of the universe as a whole.

HW: That it's made of the same stuff, and that the different parts are related, that each part is related to every other part.

WW: Each part is related to every other part by influences which we cannot describe. We don't know what these influences are, nor do we know how they're propagated. We don't even know whether the word "propagated" is applicable. But there's a basic principle that if you change one part of the universe, that affects all other parts of the universe – affects it instantaneously, I'm going to say, because I don't want to have this be longer from a distant star than it is from the sun or the moon; that the configuration of the whole of the universe – if you tinker with it, if you put a screwdriver in and diddle one little place, here, everything changes, everything is affected by that. How? We don't know.

HW: As if the whole thing were a sensitive membrane.

WW: Yes, in a way, but I'd prefer not to have any analogy, because any analogy is going to have its limitations. Now, I'm perfectly confident that nobody can possibly prove that Assumption A is valid; but I'm also perfectly confident that nobody can prove that it's false.

Chapter Twenty
WARREN WEAVER ON ASTROLOGY

Anyone who is not shocked by quantum theory has not understood it.[52]

Niels Bohr

My father and I each wrote an introduction to our book. In his "Preliminary Comments" he gave vent to his irritation with quantum physics; and even though you've heard his arguments before, I think you'll enjoy his style:

As we begin to explore together some ideas about astrology, I want first off to say something about the nature of the theories advanced by present-day physicists - especially the theoretical physicists who are seeking to develop our understanding of the ultimate constitution of matter. Among the recognized conventional categories of scientists it is probably the physicists (along with the astronomers) who have been the most severe critics of and the most vocal objectors to astrology, so they must not complain if someone chooses to examine the way they talk, and the kind of basic theories they formulate and defend.

Not much over a half century ago, physicists described atoms as composed of two types of elementary particles, elementary particles being defined as indivisible objects not composed of smaller parts. That is, an atom then consisted of a core called a "proton," around which, a little like planets around the sun, rotated attendant "electrons." This was intelligible and indeed appealing.

But as the experimental and mental tools of physics became more powerful, it has turned out that that classical picture is far too simple. By bombarding these classical "elementary" particles with missiles of great energy – fragments of other particles that have been accelerated to very high velocities – they were broken into what is, by now, a wide variety of constituent parts.

Just listen, for a moment, to a current list of what are still called "elementary particles," although it is completely obvious that the adjective "elementary" is not really applicable. As a primary distinction, particles which "feel," as the phrase goes, what is known as "the strong force" are called "hadrons," whereas particles which are not affected by the strong force are called "leptons." Particles with integral values of the spin angular momentum are called "bosons" whereas particles with half integral spin momentum are called "fermions."

In addition, some elementary particles have a property charmingly named "strangeness." And lest the reader think that these modern physicists concerned with elementary particles have used up their sense of imagination - or lost their sense of humor - it should be added that there are particles called "quarks," "anti-quarks," and "charmed quarks." In a current article in a leading scientific journal,[53] there is a discussion of quark "flavors" and quark "colors," including red, green, anti-red, anti-green, and believe it or not, "magenta quarks."

Since many physicists are obsessed with the idea that every effect must have its cause, and since the causes are frequently thought of as forces, it is fair to remind the reader that the experts in particle physics recognize the existence of four forces. Listed in order of increasing magnitude, the smallest of these is the "gravitational force." The next larger, the "electromagnetic force," is ten to the 37th power (or ten billion, billion, billion) times as large. The "weak force," the next larger, is ten to the 11th power (or one hundred billion) times as large, and the next largest, or "strong force," is ten to the 12th power (or a thousand billion) times as large.

As regards the times for which these forces act, the strong interactions are thought to persist for ten to the minus 23rd power seconds[54] (or one ten thousandth of a billionth of a billionth of a second); the electromagnetic reactions for 137 times longer, and the weak reactions for very widely varying times: as little as a billionth of a billionth of a second, or as long as fifteen minutes.

When two elementary particles interact through the agency of one of these forces, the force is supposed to be transmitted by the exchange of an intermediate particle peculiar to that force – whatever that statement means! The distances over which the electromagnetic and gravitational forces act are considered to be infinite, whereas the range of activity of the strong force is about ten to the minus 23rd power of a centimeter (or one ten thousandth of a billionth of a billionth of a centimeter), and the range of the weak force is supposedly a billion times that of the strong force.

If the statements of the preceding few paragraphs, as borrowed directly from the writings of some of the recognized authorities on particle physics, appeal to the reader as almost incredible nonsense, they will have served my purpose.

The reader may recall that in *Alice Through the Looking Glass* Alice remarked, "One can't believe impossible things," and that the White Queen countered that when she was Alice's age she practiced a half hour a day, and sometimes believed as many as six impossible things before breakfast.

But when physicists say that the ordinary solid matter of our daily experience - the matter of which trucks and bridges are constructed - is composed of elementary particles such as those listed above, then it seems a little inappropriate for these physicists to say that all the ideas encountered in astrological lore are crazy and unbelievable.

As a wholly reasonable reaction, many persons and presumably all orthodox scientists find it difficult indeed to conceive of any way in which the configuration of the planets at the time of a baby's birth can affect the personality traits and talents of that child. What conceivable "forces" could bring that about?

However, to some scientists, at least, it may seem reasonable to assume and to believe (Assumption A) that the entire universe is an interrelated unit, every part of which is related to and affected by every other part, whether or not we have a reassuring name or any description of the "force" which carries out the effect.[55]

The universality of this assumption is entirely comparable to that of the universal law of gravitation. In the classical scientific instance we have a name for the law, and a neat and simple formula for its magnitude. It may, however, be well to remember that in the case of gravity the name and formula are *all* that we have. Not even Einstein, popularly supposed to have "explained" gravity, had any basic and understandable explanation for *why* there should be such a force.

In his *Lectures on Physics*, in a section labeled "What is Gravity?," the world-famous theoretical physicist Richard Feynman writes:

But is this such a simple law? What about the machinery of it? All we have done is to describe *how* the earth moves around the sun; but we have not said *what makes it go*. Newton made no hypotheses about this; he was satisfied to find *what* it did without getting into the machinery of it. *No one has since given any machinery.*[56]

Thus at the moment of birth – i.e., the moment when a human being becomes a full-fledged member of that overall unity – the whole of the universe, however distant and in however unimaginable a manner, may exert some influence on that new member of the overall unity. What forms that influence may take, and how that influence may be correlated with certain aspects of the overall unity (in particular, with the geometrical configuration of certain celestial bodies at the moment of birth) will be discussed in later chapters of this book.

I use the term "influence" of the overall unity of the universe rather than the "force" exerted by the overall unity, because the word "force" has far too specific an atmosphere of deterministic sufficiency, whereas "influence" is consistent with the preferable notion that the cosmic unity may have a partial but not necessarily a complete and inescapably deterministic effect.

There would seem to be a second assumption, (B), which is basic to astrological thought. This is the assumption that the geometric configurations of the sun and moon and the eight planets (Mercury, Venus, Mars, Jupiter, Saturn, Uranus, Neptune, and Pluto) at the moment of birth can be correlated with certain personality traits and certain talents of the individual in question.

Whereas Assumption A appeals to me as a scientist as natural, impressive, and in fact similar to certain of the grand generalizations of classical science, Assumption B seems to me curious and improbable.

That, however, is no reason for refusing to put Assumption B to specific test. And Assumption B has one, to me, very impressive asset. It has undoubtedly been believed by many thousands, if not millions, of individuals over many centuries of time. That persistence in human experience is not to be disregarded and, in my judgment, is a powerful argument for keeping an open mind and a willingness to consider properly arranged and controlled tests.

Chapter Twenty One
CHANGES

The great underlying, and essentially unprovable, assumption on which all of science is based is that nature is orderly. . ... We keep, in science, getting a more and more sophisticated view of our essential ignorance.[57]

Warren Weaver

My father was eighty-two that fall and shortly after we started to work on our book his health took a turn for the worse. In the last pages he wrote out for me his beautiful, impeccable handwriting had become tiny, cramped, and almost illegible.

And just then, Knopf offered me a major translating job: Philippe Ariès' magisterial study of ideas and customs relating to death in the Christian West, *l'Homme devant la mort*. My father was concerned about my ability to meet the mortgage payments on my house, and he urged me to accept their offer.

I could tell that he was very tired. Perhaps, too, he was having second thoughts about jeopardizing his reputation in the scientific community. His suggestion to his old friend Gerry Piel, editor of the *Scientific American*, that they devote space to arguments for and against astrology - had fallen on deaf ears. I don't know what was going on in his mind. All I know is that the book we planned to write together was postponed indefinitely.

When Knopf sent me the manuscript of the Ariès book I almost wept: it was over 1200 pages. It would take me at least a year to complete my translation. (It took two.) I think I knew then that my father's book – our book – was not to be.

Back in Woodstock I rolled up my sleeves and went to work on the Ariès, which turned out to be one of the most fascinating books of my translating career, and the only one besides the Artaud in which I got to translate poetry.

But I did not turn my back on astrology. I continued to do charts in my spare time; and when the local astronomer (who could not resist taking pot shots at astrology in his weekly column in the *Woodstock Times*) launched a full-scale attack on what he dismissed as "pseudo-science," I joined my colleagues in what became a long, impassioned – and occasionally acrimonious – debate.

In the late seventies there was much talk among astrologers and other prognosticators of a "grand alignment" of the planets in the early 1980s that some believed would herald catastrophic events on Earth. As I recall, I was not particularly of this persuasion myself, so when Bob Berman began attacking astrology in his column "The Night Sky," I let my friend and fellow astrologer Allan Edmands handle the defense and held my fire.

Berman's first argument against astrology was very much like my father's initial objection:

Here on Earth, we're most strongly affected by the gravity of the moon, because it's so very nearby. . . .. Why should anyone worry about the movements of objects with less than one-millionth as much "pull"? Answer: because it sounds far-out and plausible to those who know little about the real solar system. Because it sells magazines and newspapers.

As you can see, his tone was very different from my father's. Berman went on:

300 *astronomers* and scientists signed a petition a few years ago condemning astrology, as currently practiced, as mischievous hocus-pocus.

Allan pointed out:

That 186 (not 300, as he asserts) scientists signed a statement in the *Humanist* magazine last fall condemning astrology as hocus pocus does not disprove astrology! Over 100 other scientists who were approached with the statement refused to sign on the grounds that they didn't know enough about the subject to make a judgment.

Berman said that his main objection to astrology was

. . . the use of astronomical or scientific jargon . . . to justify something which has absolutely no scientific basis. For despite its popularity, its elaborate tables and scientific-sounding terminology, astrology is based on no science whatsoever. No conclusive studies of thousands of people, with control groups. No experimentation. No research that could be called such.

At this point I could no longer remain silent, so I jumped into the fray, armed with the studies that had caught the attention of my father: the work of John Nelson, Vernon Clark, and Michel Gauquelin. I recycled some of the historical material in my introduction to the book we had shelved, and I concluded:

My honorable opponent seems taken with the image of himself as a hero, the man who had the guts to attack astrology in Woodstock, that hotbed of the occult. History will show that the real heroes in this debate are the men and women who have had the courage to investigate phenomena that lend support to astrology in spite of the contempt and ridicule of the scientific establishment.

My colleagues Allan Edmands, Dodie Edmands, and Mary Orser all manned the barricades with such eloquence and vigor that Bob Berman (who had, after all, fired the first shot) began calling for an end to the debate, declaring it "stuporously silly" and insisting that "if this astrology thing goes on much longer, I suspect our readers may start collapsing from boredom."

But the editors of the *Woodstock Times* knew otherwise. Controversy sells papers, and one of the staff writers, Jim Reed, had the brilliant idea of asking five local astrologers to try their hand at a "blind reading" of the chart of an unknown person, as an experiment. This person turned out to be – surprise! – none other than Bob Berman.

Allan and I both suspected that the birth data we were given was Berman's: the time and place were right, the chart "looked" like him, and it was just the sort of stunt the *Woodstock Times* would pull. But of course I didn't dare say anything that wasn't there in the chart, in case Allan and I were completely off base.

The strongest clues were the prominent Mercury, often found in the charts of writers and/or teachers – Berman's astronomy classes were popular – and Saturn in close conjunction to the Ascendant, an aspect that Gauquelin had found in the charts of scientists with a frequency greater than chance. That Mars was also hugging the Ascendant within half a degree suggested a fighter, which was certainly the case. A feisty scientist who taught and wrote!

My thumbnail sketch was so accurate that my friend Bob Reynolds congratulated me for "knocking 'em dead in the paper." It was fun, but I always felt a little guilty about my extra-astrological suspicions regarding the identity of the subject. The "experiment" could hardly qualify as science.

Although I was back in Woodstock – or rather, in my house in Riverby, about eight miles from town – I felt very connected to my failing father and visited my parents in Connecticut every two or three weeks.

Dad was one of those people – like myself – for whom work is the ultimate addiction. Physically uncomfortable most of the time and unable to do much of anything – even weeding the garden exhausted him – he gradually sank into depression. I remember one day I came upon him in his shop, struggling to insert a nail into a picture hook. "My old hands don't work," he said, half to himself. It broke my heart to hear my father the carpenter say that.

And another day when we had ordered Chinese food and Mother and I were opening our fortune cookies and reading our fortunes aloud he said bitterly, "I know what my fortune is." When I asked him how he was one morning he said he felt like an endangered species.

He continued to rescue me whenever I was unable to meet my expenses. I started to think maybe he was just holding onto life for my sake, because he didn't think I could make it on my own.

The following exchange, recorded in my journal, is typical of those days. Dad had generously offered to pay for body work on my car following a close encounter with a deer. I was leaving to go back to Woodstock:

HW: Private Weaver requesting leave, sir.
WW: Well, they can lick one or the other of us, but they can't beat the combination. You're a good daughter.
HW: You're a good father.

WW: We're a good team.

HW: You bailed me out again, Pop.

WW: What are fathers for? Who better?

HW: Some day you should be a little tougher on me.

WW: Drive carefully. Not too fast!

HW: That deer hit me. I didn't hit her: I hope that's understood!

WW: Do you know what my father said when he and another man almost sideswiped each other? "He was much closer to me than I was to him!"

In the summer of '77 I took up with a man with whom I was to have the most intense and dramatic – and perhaps the most painful – love affair of my life. The history of our ups and downs is so extreme it's almost comical. I just read through those journals again after thirty-two years, and I didn't know whether to laugh or cry. I did both.

I think the reason it was so painful – aside from the fact that I wanted a commitment and he didn't, but warmed up whenever I tried to distance myself – was that in so many ways we were good together. It was such a near miss.

I've always been drawn to foreigners, but Germans are a special case. My father's ancestors were German and he grew up bilingual. I've always been proud of my German blood. I've always loved the sound of the language – the language of Mozart and Beethoven and Goethe – and longed to be able to speak it myself someday. This man was born in Germany in World War II and grew up speaking German, English, and French. He was irresistibly handsome, had a delicious accent, spoke idiomatic French, and had a beautiful tenor voice, a brilliant mind, and a sense of humor (and appreciated mine). He was a talented artist and photographer, was into meditation, and played the flute. The sex ranged from friendly to spectacular, depending on our mood.

In short, I was a goner. He was also a little nuts and as my father said after meeting him, "You have a weakness for nuts." You'd see what Dad meant if I described his work, but I can't do that because it might blow his cover, and he's a *very* private person. Let's just say that his work as an artist is on the cutting edge and that at the time our affair began he had no visible means of support. I'll call him Hans.

Hans spent three months living in the cabin on my land. My previous tenant had become ill and she had left her dog Jessie in my care when she went home to Connecticut to recover. To everyone's surprise Jessie was pregnant, and the night Hans and I became lovers her puppies were born.

There were seven pups in the litter: six females and one male. I found homes for three of the females and took the other three to a no-kill shelter in Kingston, but the male was special. I adopted him and named him Pluto.

I haven't had much experience doing horoscopes for pets, mainly because most of my pets have been rescues where no birth data has been available. But Pluto was born under my cabin the same night Hans and I embarked on our stormy affair, so in effect Pluto and the relationship had the same chart. I didn't actually cast a chart for Pluto or for us, but I did notice that major events in the dog's life paralleled the rocky course of our affair. The same day Hans and I had a knockdown drag-out fight, Pluto killed a woodchuck. The day of another painful confrontation, Pluto tangled with a porcupine. Pluto was an outdoor dog: he was born outdoors, and he hated fences. Our relationship was wild and a major issue was Hans's fear of commitment: Don't fence me in!

Riverby was wild, too. I had chosen it for that reason; but it wasn't long before I discovered that I had bitten off more than I could chew. My driveway was a disaster. Steep and winding, it became impassible for months once the Catskill winter set in. Sometimes I was trapped in the house for days because my VW Bug Sadie couldn't make it up to the road. When I did manage to get out to town and back I had to park on Middle Way and walk down a treacherous slope which, with arms full of groceries, was downright dangerous.

The roads in Riverby were bad too and Sadie didn't do well in new-fallen snow, ice, or slush. When I went to a winter party I never knew whether I was going to be able to get home or not. One night the week before Christmas I was driving home from a party in a blizzard. It was about 12:30 am and the snow was coming down fast. I got partway up Abbey Road (the main road of Riverby was named after the Beatles album) when I realized that Sadie wasn't going to make it, and I had to back down again. I tried again, but it was no use: unplowed fresh snow on top of ice. At the bottom of the hill I tried to turn around but there wasn't room, and when I tried backing up clear to the Wittenberg Mount Tremper Road (up a slight rise) I overworked the transmission and the battery died.

I was alone in a dead car. It was now 1:00 am and snowing steadily.

I got out of the car and knocked on the doors of a couple of houses, hoping to wake someone up to make a phone call, but no one responded. I ran out on the road and started flagging down cars, doing Nixon impressions with my arms. The first car ignored me. The second one contained a female acquaintance.

She asked me, "Where do you want to go?"

I said, "Anywhere! Just get me out of this white stuff." I ended up spending the night in her house.

I always felt lucky that the first car that stopped for me did not contain Jack the Ripper or any other serial killer. A lone woman hitchhiking at 1:00 am is not recommended. It was two days before my abandoned car could be dug out and jumped and I could go home.

I got bailed out of that one, but I think that's when I realized that my situation living in Riverby – the isolation and the iffy roads and my bad driveway – was not tenable. The house into which my father and I had put so much love and energy – and in which he had invested so much money – was in the wrong place. It was not the house of my old age.

Chapter Twenty Two
LACRIMAE RERUM

Father Breath once more farewell
Birth you gave was no thing ill
My heart is still, as time will tell.[58]

Allen Ginsberg

In May 1978 I wrote my father a very difficult letter. In it I described my experience getting stranded in a blizzard. I told Dad that I didn't want to spend another winter in that house. I told him how bad I felt that his incredibly generous gift had somehow turned out to be a burden, but I felt I had to sell the house.

At first there was silence on his end, but after a few days he called. He said he'd been upset but had thought it over and now realized I had made the right decision. He said he was behind me a hundred percent. (How many times had I heard him say that?) We agreed there was no point focusing on the negative.

Mother on my Riverby adventure: "It was a little like the new pope – he was only in for thirty-four days!"

Some time that summer Pluto had a run-in with a porcupine and came to me with a nose full of quills. The vet had to put him out with Nembutol in order to remove them.

When it comes to porcupines, there are two kinds of dogs. The first kind learn their lesson after one encounter. The second kind are out for revenge. Unfortunately, Pluto belonged to the second category.

After a second encounter I took him to New Milford to stay with my parents for a while. When my mother fell and broke her arm I moved in myself to help take care of them.

My parents fell in love with Pluto, and once he'd made a hole in the screen door so he could come and go as he pleased, he adjusted to life in Connecticut.

My father was particularly fond of the dog. He wasn't feeling well in those days and had a hard time getting out of bed in the morning. But Pluto would race into his room and lick his hand and Dad would smile and find the courage to face his day.

But Pluto was an unaltered male, and he started hanging out in neighbor's yards in search of romance. There were complaints, and Mother got a call from the police. So I took Pluto back to Riverby and left him with my neighbors there until I could find a permanent home for him.

I put an ad in the *Woodstock Times* with a picture of Pluto. I put the house on the market. And after a bumpy summer with Hans, I decided to stop seeing him.

So many endings! In October I sold the house and rented a little cottage within walking distance of Woodstock.

One morning later that month Mother called to tell me that my father was in New Milford Hospital. He had fallen in the bathroom and hurt his back. I rushed to Connecticut. I found Dad flat on his back – it hurt him to be raised – and barely able to speak. He was refusing to eat, wanted no phone or TV.

Now that her arm was healed Mother didn't need me, so I went back to Woodstock for a few days. When I stopped at the hospital to say goodbye to Dad, I could tell he recognized me, but he didn't respond.

I asked him if he had any messages for Pluto. He smiled then and said, "Tell him to stay away from porcupines."

For the next few weeks I went back and forth between Woodstock and New Milford. My father was failing fast. I knew that he was not going to come home from the hospital, and I wanted to spend as much time with him as possible.

All of his life, Warren Weaver had struggled to reconcile scientific materialism with the faith of his fathers or, as he put it, science and religion. Occasionally he had succeeded. His Scientist's Prayer, written in the forties, is imbued with a sense of God's presence within the creation. But this sense of immanence was not something he could sustain. Somehow his philosophy failed him at the end. It was not enough to cope with his own death, which he perceived as the total annihilation of his being.

There is a passage from the *Rubaiyat* which he loved and which evokes this pessimistic view:

Myself when young did eagerly frequent
Doctor and saint and heard great argument
About it and about, but evermore
Came out by that same door wherein I went

And then the seeds of wisdom did I sow
Watered with mine own hand to make it grow
And all the wisdom that I reaped was:
We come like water, and like wind we go.

To face death with a philosophy like that requires an incredible act of courage. I didn't believe that death was the end. My experience had brought me to a belief in reincarnation and the continuing life of an evolving entity that satisfied my reason and nourished my spirit. I wanted to breathe that belief around Dad's bed. I wanted to be with him when he died. I wanted to experience those precious last moments as fully as possible.

On November sixteenth my mother gave the hospital permission to take my father off intravenous feeding. She knew that he did not want to be kept alive artificially, as she put it.

On that same day a man called in response to my ad in the *Woodstock Times* about a home for Pluto. On the eighteenth I went to Riverby to meet the man who answered my ad and introduce him to Pluto. He was looking for a companion for his female shepherd mix, Jessica. The meeting was a success: Pluto fell in love with Jessica on sight, and got right into the man's car.

I drove back to Connecticut and went straight to the hospital. I got there after visitors' hours were over and headed straight for Dad's room.

Dad's nurse Michele told me he had stopped asking for pain medication but he had also stopped speaking. She said it took all his energy just to breathe. He looked peaceful. Michele said he liked to be turned slowly and softly, and noticed the difference. She had asked him two nights ago whether he was comfortable with the way she was doing it and he stroked her hand twice. That made me happy because it meant he liked her.

Michele said, "They take in a lot at this stage, even though they can't respond." I remembered reading that in Kübler-Ross. I also remembered reading that immediately before death, the need for analgesics decreases.

On November twentieth Dad looked right at me and even tried to say my name, and I knew that he knew me. I gave him a big smile. I felt as if he almost returned the smile. I felt his love. But I also felt an unutterable, unbearable sadness, that he had to leave the ones he loved. I realized that I was grieving not just for my loss, but for his; not just because I was having to say goodbye to him, but because he, who loved us so much, was having to say goodbye to us.

I sang to him: lullabies, hymns – anything German I could remember, hoping he'd think it was his mother.

Those days I was only really at peace when I was with Dad.

When I was in his hospital room – whether I was sewing or praying or just sitting with my father – I concentrated on radiating my belief in the world of spirit. After a while the room itself changed. I began to feel that there were presences – angels, if you will – in that room. I almost imagined that I could hear their wings.

My father's eyes – those lovely hazel eyes – had taken on a look full of serenity and space, a look that spoke of planets and galaxies.

Years before, I had had a dream in which my father and I were standing on the deck of a ship and the sun was coming toward us. We knew that it was the end of the world and that the Earth would be burned up and everyone would die; but we were too excited at the momentous event that was happening to be afraid or sad. We were glad that we were together at such an important time.

Being with my father in that room had some of the quality of that dream. Something terrible was happening, but at least we were going through it together.

Room 210 became a place of love and light, a sacred space, a magnet I was drawn to – a place of joy. It was the only place I wanted to be. I could hardly tear myself away.

As soon as I left that room I experienced the loss to come. In the elevator, in the parking lot, I fought back the tears. Driving up Second Hill past the cemetery, I sobbed at the wheel of my car.

The family gathered for Thanksgiving. My nieces Annie and Melissa arrived from New York and Vermont and my brother and his wife Marianne from D.C. Nip was ambivalent about seeing Dad but I talked him into coming to the hospital with me, and later he thanked me for it.

On the morning of Thanksgiving Day Mother and I had a little confrontation. We were sitting in Dad's room in the hospital, she by the head of his bed and I at the foot. She began planning his service. I had told her that Kübler-Ross says even people in a coma take in a great deal of what's going on around them.

I asked her to stop talking about it. She said, "He can't hear me!"

I said, "We don't know that!"

Then when I was kissing him goodbye she jabbed me in the ribs and said to hurry up. So I stood aside and told her to go first. But she wouldn't do that, she just wanted me to finish up so we wouldn't get off schedule. When I objected she said, "He's my husband!" and I countered with "He's my father!"

Then I laughed and told Dad, "We're fighting over you!" and he seemed to smile.

Mother told me that his favorite hymns were "O Love that wilt not let me go" and "Dear Lord and Father of Mankind." So at home that afternoon I put my tape recorder on the piano and taped those two hymns, plus my favorite, "Abide with me."

That night I took my sewing and my tape recorder to his hospital room and sat beside him, singing the hymns I had recorded.

I felt as if his whole room was full of light. I was happy to be there.

The next morning at 8:00 a.m., Mother walked into the room at home where I was sleeping and I knew before she said the words, "It's over." I got out of bed and held her while she wept – something I'd rarely, if ever, seen.

My nieces and I hurried to the hospital to say our last goodbyes.

In Room 210 I had a few moments alone with my father. I kissed him and put my hand on his forehead – that noble intelligent brow, the skin taut and dry from his long fast. I touched his shoulders – nothing but bone – trying to memorize the feel of his body as one does with a lover who is leaving. With the scissors I had brought from my sewing kit, I cut a lock of his hair.

I told Dad how much I loved him. I told him that he was the best father in the world, and that I wanted to make him proud of me. I promised him I would finish our book.

Back home my brother was working on Dad's obituary for *The New York Times*. Dad's death coincided with the story of the mass suicide of 800 Americans in Guyana. Nip said that his job was almost impossible because of the amount of material to choose from. He didn't actually write the story, but phoned in the material for the re-write man. This turned out to be Pranay Gupte, who was actually the man the *Times* sent to cover the Guyana story.

The *Times* obituary quoted my father as saying, "The ways a scientist and a poet go at a problem: technically they are different, but their basic goal is the same: to sense some simple beauty and unity in the complexities of the world."

Dad's service was held in the little chapel behind the First Congregational Church in New Milford. Victor Brown, a retired minister and a great friend of the family, gave the eulogy.

The baritone who sang the hymns must have been moved to tears, because he had serious breathing problems. I had to stop praying for Dad and concentrate on helping him get to the end of each line.

But nothing could have been more perfect than Victor Brown's reading of Dad's Scientist's Prayer.

After it was over Mother said, "Victor was twice too long and glorified Daddy, and I was dying to go to the bathroom for the last forty minutes!"

That week flowers and letters of condolence poured in from all over the world. Mother got one from the president of the Sloan Foundation crediting Dad with "achieving immorality" for his accomplishments – a classic boner my father would have loved.

The man who sold us the plot at New Milford Cemetery said we'd have to hurry to get Dad buried before the ground froze.

When the funeral director delivered Dad's ashes I took them to his workshop and opened the package. I felt strangely guilty, as if it were an act of disrespect – even violation – and asked Dad to forgive me my curiosity. As a scientist, I felt he would understand.

Mother said the ashes "weren't Daddy," and I agreed; but I pointed out that if she really believed that, she'd throw them in the garbage. I told Mother they're like a house you once lived in. You wouldn't want it to fall into disrepair.

I opened the cardboard carton and pried the cover off the metal cannister with one of Dad's screwdrivers. Inside were the bone-white pieces of calciferous material – some porous, some sharp hard splinters, some even colorful – that had formed the armature of my father's body. I put a handful in a paper bag to take home with me for my altar.

I wrapped the cannister in five layers of white tissue paper and tied it with a blue cord I found in the Christmas wrapping box. Then I went outside and found a fallen pine branch, cut off a sprig, and stuck that under the cord. As Mother and I left to go to the cemetery, I noticed four of Pluto's balls I had put on the hall chest to take to his new owner. Pluto had loved Dad, and Dad Pluto, and Pluto had given him some of his last good moments. The blue ball seemed to want to go with Dad, so I took it along.

When I drove to the grave site that afternoon the gravedigger was there waiting in his truck. I saw the inevitable green cloth over the hole and the dirt he had dug out. Mother said she'd stay in the car. The man asked me whether I wanted him to put the box in the ground or preferred to do it myself, and whether my mother and I would like to say a little prayer. I told him I'd do it for her. I asked him to leave me alone for a minute.

I knelt on the damp ground and lowered the box into the hole. The gravedigger had said there were three or four inches of frost and he had to cut through it with an ax.

The box looked pretty with its pine branch. It look like a Christmas present I was giving to the Earth. I took Pluto's blue ball out of my pocket and added it, and it looked right too. Then I put my hand under the green rug and took a handful of the rich brown earth, thinking as I did so how much my father the gardener had loved the feel of the earth and how many times he had made the same gesture himself.

I thanked God for giving my father a peaceful death. On the way home Mother said we'd come at Easter and plant a flowering almond, if the ground was willing.

I wrote a piece called "Saying Goodbye to Dad" for the *Woodstock Times* and gave it to Geddy along with a picture I took of Dad in the garden in 1970 and his Scientist's Prayer. Geddy made it the lead article in the Christmas issue. All of Woodstock wept over my father.

He was the best man I've ever known. Jack Kerouac was right: he was the love of my life.

Over the years since his death I remembered my promise to write our book myself. Or if not exactly that book at least some other book that would incorporate our dialogue about astrology and science. My grief at losing him has been inseparable from my gratitude for that dialogue, for the grace of our coming together at the end of his life, and for those rare qualities of his of humility, integrity, and open-mindedness that made possible such an improbable meeting of the minds.

Two years after Dad died McGraw-Hill hired me to translate a small paperback *dictionnaire de l'astrologie* published by Librairie Larousse. The book turned out to be very poor work. It abounded in misinformation, including elementary errors in astronomy, and lacked any reference to well-known French astrologers, much less to the important work of the Gauquelins. I asked my editor at McGraw-Hill for permission to revise and expand, and received *carte blanche*.

I hired Woodstock astrologer Allan Edmands as my technical assistant, and we went to work, correcting the author's mistakes, cutting drastically, rewriting and expanding most of the remaining articles, and adding over three hundred articles of our own and sixty diagrams. I also enlisted, and got Larousse to subsidize, the help of three distinguished astrologers, Charles Emerson and Robert Hand in America and Charles Harvey in England, as consulting and contributing editors. Each of these men went over the manuscript carefully and made many criticisms, additions, and suggestions, often quite detailed. We incorporated their material and did as much further research as we could in the time allotted to us.

All this was done without a royalty, since theoretically I was just the translator and Allan a researcher. My editor felt guilty enough to give me credit as general editor and co-author, along with Allan, on the title page; let us sign a few of the articles we wrote; and let me write a preface placing the book in its historical context and telling a little about how it came to be – without, of course, hinting that the French text was in any way inadequate.

I dedicated my work on the *Larousse Encyclopedia of Astrology* to my father, "the late Warren Weaver, a scientist who listened and who at the end of his life, found the courage to open his mind to astrology."

How does astrology work? We don't know. But the answer may have something to do with light.

"Translation of light" is a technical term borrowed from the Arabs, but it is also a good overall description of what astrology does. In effect, an astrologer is a translator. An astrologer translates the light of the Sun, the Moon, and the planets into ordinary language, decodes the meaning of their motions. If the planets are simply signs, or time keepers, then there is no need to find a mechanism of influence. If, on the other hand, they operate by some kind of physical influence, then the mechanism is a mystery that remains to be solved.

It may be that, as with the wave/particle dilemma of physics, both are true: that the planets and the luminaries function as both signs and causes, and the fact that a photon of light does not diminish with distance may be a clue. Signs and causes, like waves and particles, may be two ways of looking at the same phenomenon. Two hemispheres of the brain: two kinds of attention. Astrology and science: two languages for observing the same reality.

Whatever the answer, we can be certain about one thing: more light needs to be cast on the subject of astrology. Perhaps in the so-called Age of Aquarius, astrology will once again come into its own. Perhaps one day at Warren Weaver Hall, the mathematics building at New York University, astrology will be taught again along with mathematics. Perhaps one day

Warren Weaver, twentieth-century *mathematicus*, pioneer in molecular biology, computers, and machine translation, will be honored as a pioneer in the reintegration of astrology and science.

APPENDIX NO. 1

THE WEAVERS AND THEIR STARS

Chart comparison (is) the comparative study of two or more charts of individuals, nations, corporations, and so on, for the purpose of determining compatibility, improving communication, or illuminating problems within a personal, political, or professional relationship. In chart comparison particular attention is paid to major aspects between charts, especially conjunctions and oppositions, and especially those between Sun, Moon, and Ascendant. In addition . . . major configurations, heavily afflicted planets, and other factors important in interpreting single charts are also brought into play.[59]

Larousse Encyclopedia of Astrology

As we have seen, chart comparison is also known as "synastry," from the Greek *syn* (with, together) and *aster* (star). When the charts involved are those of members of the same family, the issue of heredity arises.

As we have also seen, the Gauquelins came upon statistical evidence of what they called planetary heredity. For example:

Analysis of the birth charts of 30,000 parents and children . . . revealed that children of parents with Moon, Venus, Mars, Jupiter, or Saturn angular [rising or culminating] tended to have those planets angular too, with a frequency whose probability of occurring by chance was 1 in 100,000. When both parents had the same planet angular, the effect was doubled, which accords with the laws of genetics.[60]

There are other ways in which comparison of the birth charts of family members hints strongly of heredity. Although the evidence is anecdotal rather than statistical, many astrologers have noticed that signs of the zodiac tend to "run in the family," and that factors like rising signs and major configurations seem to be inherited as well.

Let's keep these ideas in mind as we examine the birth charts of the Weavers.

It is impossible to cast an accurate birth chart without an accurate birth time. In my father's case, this information was unavailable to me during his lifetime. But shortly after he died, I had a stroke of luck.

I was visiting my mother in New Milford. I slept in the room we called the Annex, where my father had kept his files, including old letters and family history, and some of the less technical books in his library.

One night I woke up at 3 a.m. Unable to get back to sleep, I was poking around in my father's book shelves, when a book fell on my head. This book was not interesting, but it startled me and there on the shelf in front of me I saw a small volume I had never noticed before. I pulled it out. The size and shape of the autograph books we used to pass around in grammar school, it was obviously very old. The cover depicted two figures in nineteenth-century dress: an elderly gentleman in spectacles weighing a baby in a pair of scales and a kneeling maid helping to support the precious bundle. The title, in red capitals, was "The Baby's Journal."

On the first page, in handwriting so faded as to be almost illegible, I read: "Baby Warren Born Tuesday morning 6:30 July 17, 1894. A fat dimpled darling. Weighed eight and three fourths pounds. Cried lustily at once; sucked his thumb before being washed. . ." and so on.

It was my father's baby book! Now I was wide awake and so excited that I knew there was no possibility of sleeping until I had cast his birth chart. Luckily, I had brought my slide rule, ephemeris, and table of houses with me. And so, at last, I had my father's chart.

I noted with interest that the time of his birth was not that different from the time I had arrived at when I had guessed that he might have Virgo rising. The 6:30 a.m. birth time gave a rising sign of Leo, the sign that immediately precedes Virgo in the zodiac.

Weaver: TRANSLATION OF LIGHT

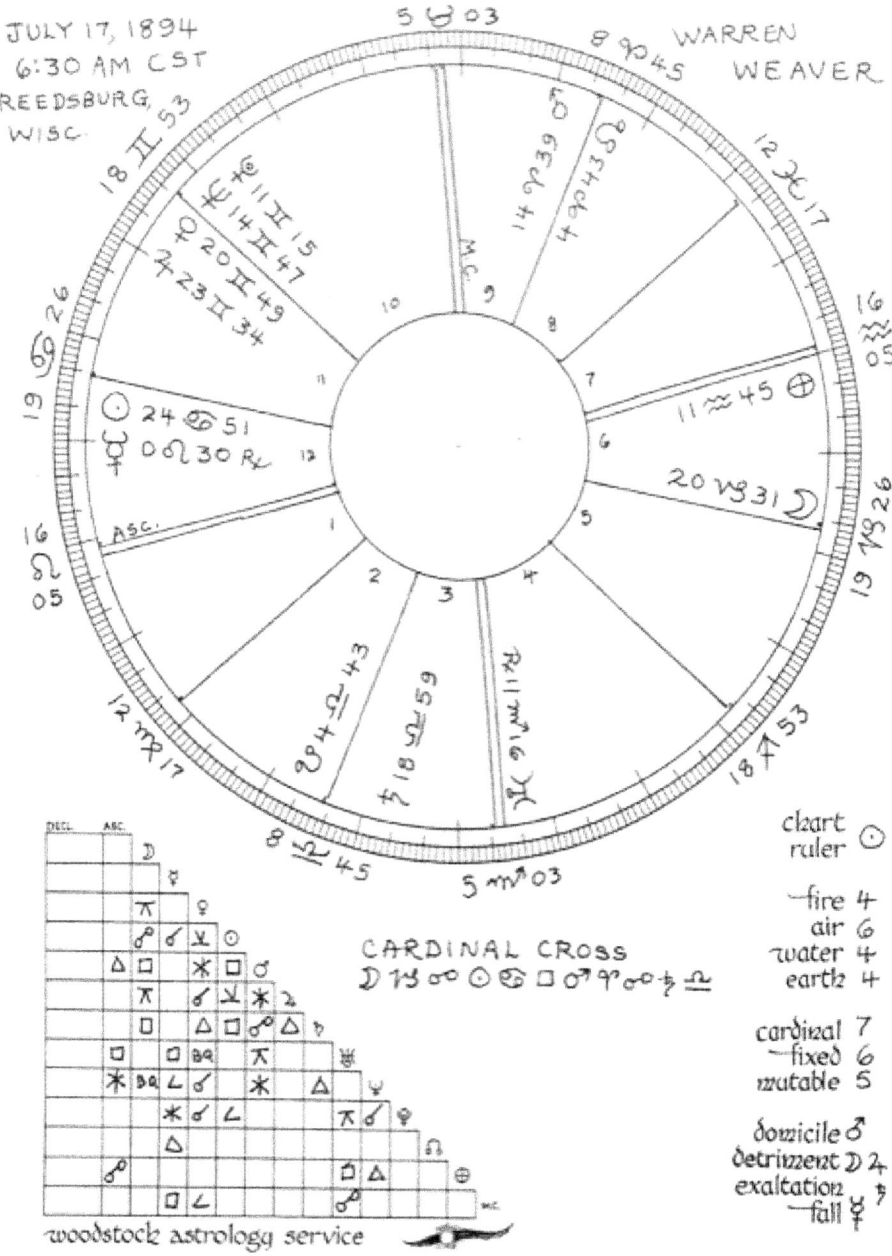

Leo people are born leaders. They belong at the center of things, where they enjoy giving pleasure and receiving attention. Leo rising made sense for Dad, who was a great raconteur and very much at home in front of an audience.

The 6:30 a.m. birth time puts Dad's Cancer Sun in the twelfth house, a "dark" (not particularly public) house which rules schools, hospitals, research, and activities that take place behind the scenes. Cancers are the caregivers of the zodiac, and foundations are sheltered from the noise and pressures of the marketplace. My father's life work of allocating funds for a foundation that supports research in science and medicine is a perfect expression of that placement for his Sun.

My father was born at the Full Moon, which always provides an extra burst of energy to a birth chart. William Blake said, "Without contraries, no progression." At the Full Moon, the Sun and Moon – the two luminaries – are in opposition. On the one hand, there can be a sense of inner conflict, of two opposing energies in the psyche; but on the other hand, there is literally more light on the path, and more opportunities for integrating desire (the Moon) with action (the Sun).

Dad's Full Moon combines the nurturing, empathic qualities of his Cancer Sun with the sober, organizing qualities of his Capricorn Moon. The Moon rules childhood, and Capricorn is not an easy placement for the Moon; in *Scene of Change*, as we have seen, my father writes that his childhood was not particularly happy and that he suspects that "childhood is often, and perhaps even characteristically, a troubled and unhappy time."

We have seen that Dad's Sun and Moon are part of a relatively rare major configuration called a Grand Cross which, according to *Larousse*, "indicates a determined, dynamic, forceful personality with a strong sense of purpose or destiny – the 'self-made man or woman.' "[61]

A cross involves four planets: two oppositions (planets 180 degrees apart) that are in square (90-degree) aspect to one another, forming a cross. A cross can be either Cardinal, Mutable, or Fixed. The Cardinal Cross is the most powerful, as the Cardinal signs are all about initiating action.

Dad's is a heroic example of the Cardinal Cross, for it involves an opposition of the Sun and Moon in square to an opposition of Mars and Saturn. Many astrologers consider the so-called "hard" aspects (square and opposition) of Mars and Saturn to be the most difficult in astrology, since they combine the two planets that have traditionally been regarded as "malefic" (i.e., unlucky). And in my father's chart, Mars and Saturn are particularly powerful.

In astrology, each planet is connected with certain signs: the sign it is said to "rule"; the sign where it is said to be "exalted"; and the two signs opposite those, where it is said to be in "detriment" or in "fall." For example, Mars, the aggressive red planet, is said to rule Aries, the energetic sign the Sun enters at the vernal equinox. And Saturn, the serious planet associated with the limitations of time and space, is said to rule the practical and serious sign of Capricorn, and to be exalted in peace-loving, diplomatic Libra. My father's Mars is in Aries and his Saturn is in Libra.

With that intense Mars-Saturn opposition squaring the luminaries, Dad had a difficult start in life, but he also had enormous energy, and was known for his ability to get along with people of differing and even opposing views and to help them work together as a team.

So this is a Cardinal Cross in spades. It describes a person who is highly motivated, industrious, and ambitious, able to overcome early obstacles, and determined to do his very best work and to contribute something of value to the world.

Two signs frequently found in the charts of scientists are Capricorn, with its attraction to structure, and Gemini, with its wide-ranging curiosity. Geminis are the great communicators of the zodiac, and often have talents that lie in several fields, typically teaching, writing, translating, science, and photography.

With his stellium (three or more planets in one sign) of Venus, Jupiter, Neptune, and Pluto in Gemini, my father did all of these things, and did them well. He was a talented photographer and had his own darkroom. And while he was not himself a translator, he was fascinated by the phenomenon of translation and wrote about it at length.

According to the *Larousse Encyclopedia*, Gemini people "seem to function best when doing two things at once. Typical Geminians have hobbies that may be just as important as their profession, or may even become their profession."[62] Dad's Lewis Carroll collection began as a hobby, but ended up a second profession.

The fact that he became well known and distinguished in his field had much to do with his ability to write. Mercury, the planet of communication, is in close aspect to his MidHeaven, which is the most public place on a chart. His forte as a science writer was his gift for explaining complex ideas in simple, elegant language. A good example is his contribution to *The*

Weaver: TRANSLATION OF LIGHT

Mathematical Theory of Communication, which has been continuously in print for over seventy years. His success also had to do with the creative genius that produced the ground-breaking memorandum on machine translation. Uranus, Mercury's "higher octave," the planet that rules cutting-edge technology, is also in close aspect to his MidHeaven.

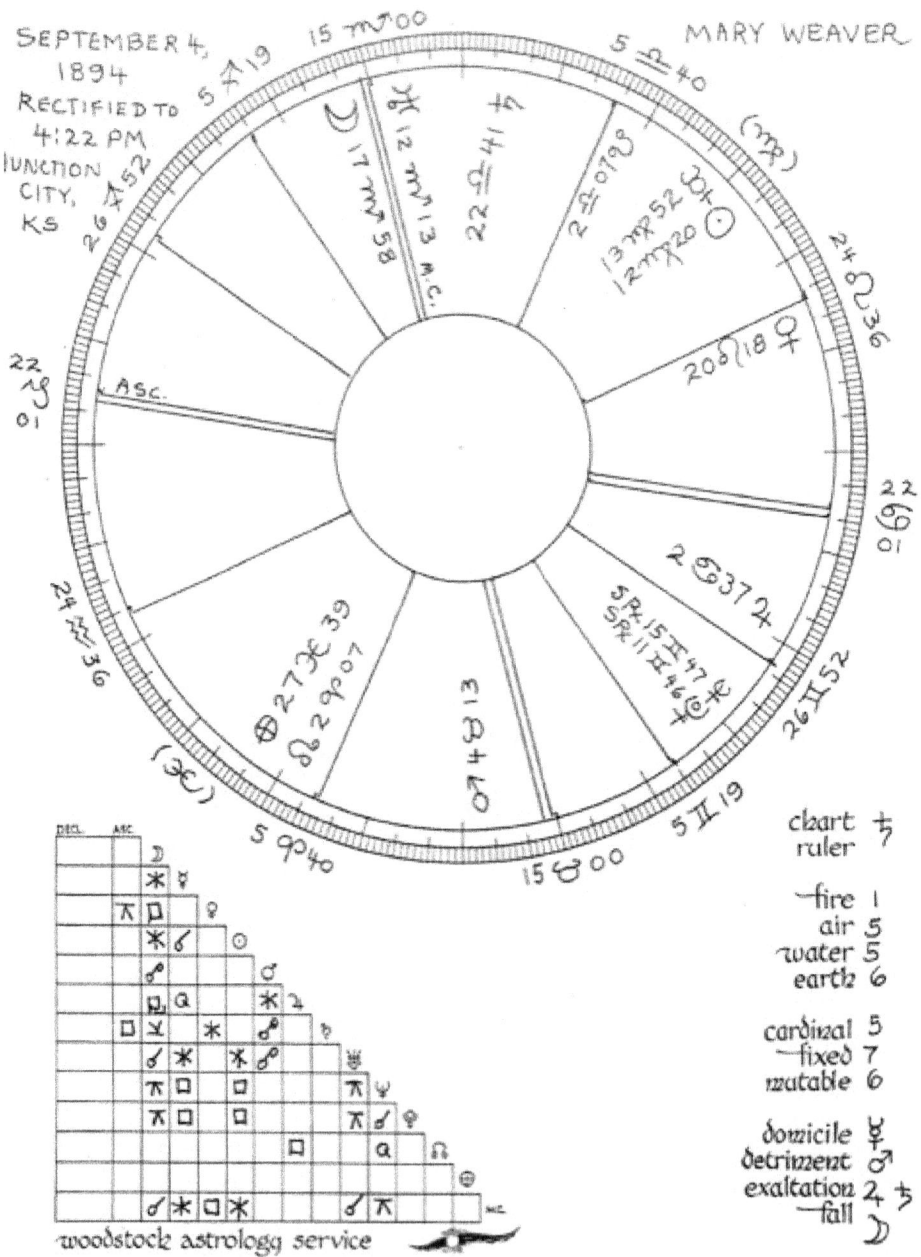

My mother was born September 4, 1894 in Junction City, Kansas. Alas, I have no way of knowing what time she was born. There was no baby book or birth certificate for Mary Hemenway – at least, none that survived – and since Grandma Hemenway died seventeen years before I began studying astrology, I had no one to ask for the information.

In the absence of a birth time astrologers sometimes resort to a chart cast for either sunrise or noon on the day the child was born. A noon chart will be no more than twelve hours off the actual time of birth. Of course, one cannot be certain of the Ascendant or the MidHeaven, or indeed any of the intermediate house cusps; and if the Moon changed sign that day (as it does every two and a half days), then there will be no way to be sure of its sign, let alone its degree of longitude. But a noon chart, like a sunrise chart, is better than nothing.

Luckily, in Mother's case the Moon was in Scorpio for all twenty-four hours of the day she was born, so we can be sure of its sign, as well as the signs and approximate degrees of the Sun and the planets.

It is a shame that we have so little information on my mother's chart, for her relationship with my father was that rare thing, a marriage made in heaven.

Mother's Sun and Mercury are conjunct in the sign of Virgo. Virgo women sometimes possess a pure beauty which, combined with the typical Virgo modesty, can be captivating, and my father fell in love with her for life. Virgo is a practical, down-to-earth sign, and Mother's Virgo Sun and Mercury provided the perfect grounding for my father's emotional, empathetic Cancer Sun.

With her Moon in the fixed water sign of Scorpio, Mother had strong emotions too, but was reserved about expressing them. She did love me, but as a child I got far more mothering from my Sun-sign Cancer father than I did from her.

Mother's Moon-Uranus conjunction in Scorpio has an opposition from Mars, a potentially explosive aspect, as it combines the willful energy of Mars with the rebelliousness of Uranus. I remember Mother telling me that Grandma Hemenway had called her "my most difficult child." So my reserved mother was really sitting on a lot of negative emotions that she was often unable to express.

With her Venus in Leo, Mother was generous, a wonderful hostess, and had an army of friends. Her Leo Venus is just three degrees away from Dad's Leo Ascendant. The Ascendant-Descendant axis rules marriage, and the Venus connection spells the kind of friendship that is based on similar tastes and values and creates a solid foundation for love.

The only other personal connection with my father that can be gleaned from Mother's noon chart is the close conjunction of her Mars in Taurus with his MidHeaven. This describes a connection that was both sensual and spiritual. Mother provided exactly the kind of energy he needed to help him give his many talents to the world. For example, it was my mother's enthusiasm for biology that helped to convince my father that biological research was the wave of the future.

Since my parents were born only seven weeks apart, their outer planets were all in the same signs, giving them the shared perspective of their generation.

Weaver: TRANSLATION OF LIGHT

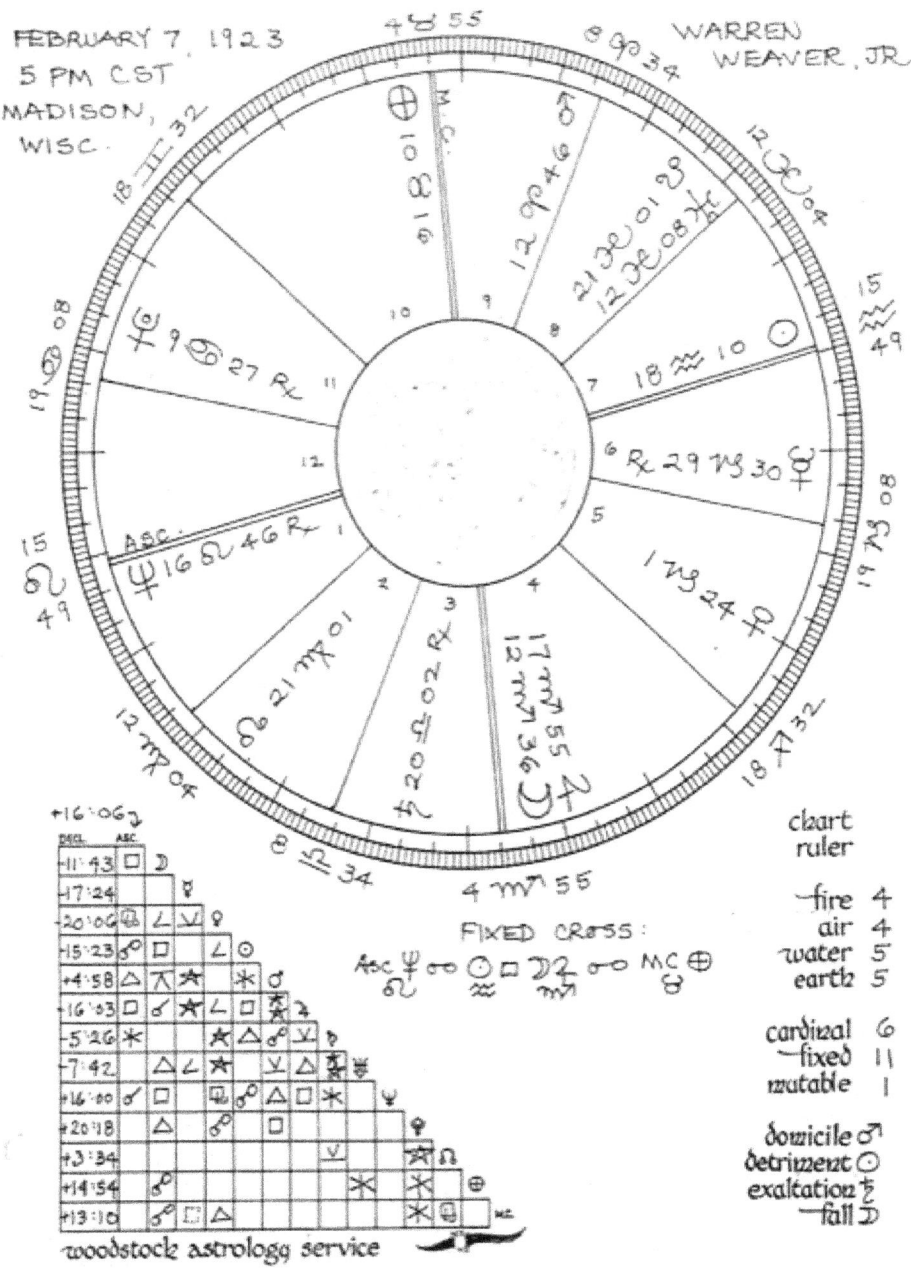

My brother, Warren Weaver, Jr., was born February 7, 1923 – at 5 p.m., according to Mother – in Madison, Wisconsin.

His Moon is in Scorpio, like our mother's, and even though we don't know the exact degree of her Moon, by consulting the ephemeris and noting the Moon's motion for the day of her birth, we do know that her Moon will fall somewhere within orb of his. (The word "orb" refers to the radius within which an aspect is considered to be potent. Since the Moon moves so rapidly, its aspects are given a rather wide orb.)

My brother had a very strong connection with our mother – much stronger than mine. The family dynamic was structured along Freudian lines: My father and I were joined at the hip; we were the ones with the "artistic temperament" who showed our feelings. My mother and my brother, with their shared Moon in secretive Scorpio, were the reserved, practical ones who kept their feelings to themselves.

My brother's Moon is not only conjunct Mother's Moon; it is in close sextile (60 degrees) aspect to her Sun and Mercury in Virgo. The sextile is an aspect of harmony, since it unites planets in elements that work well together – in this case, earth and water. The emotional intensity of the water sign Scorpio appreciates the groundedness of the earth sign Virgo. The two of them had similar temperaments, and understood each other very well.

My mother's tendency to repress her feelings may have had something to do with her Saturn in Libra, a placement she shared with her husband and passed on to her son. Libra is the sign in which Saturn is said to be exalted, and Saturn in Libra is all about judgment, and a strong sense of morality (what Freud called the superego). Everyone in the family had Saturn in Libra except black sheep rebel me.

In fact, my mother's and brother's Saturns are conjunct: Nip was born the same year that Mother was experiencing her first Saturn return, a time of increased responsibility, the astrological graduation into adulthood.

The 5 p.m. time of my brother's birth gives him Leo rising: the same rising sign as his father, and even the same degree, within only twenty-one minutes of arc. As you may remember, the zodiac has twelve signs of thirty degrees each, and each degree has sixty minutes, for a total of 21,600 minutes around the wheel. So a conjunction of only twenty-one minutes separation out of a possible 21,600 is pretty impressive. Unlike my father, I'm not an expert in probability; but clearly, the probability of that occurring by chance would be very slim.

Leos love an audience. At the annual dinner of the Gridiron Club in Washington, DC, where politicians were roasted by members of the press, my brother performed before audiences that included congressmen, senators, and even the president. He had a knack for writing satirical new words to show tunes and popular songs, and although he had claimed to be tone-deaf as a child, I remember hearing him belt out his own lyrics from the stage, right on key.

In addition to his Leo Ascendant my brother also inherited his father's Cross, but in Nip's case, it was a Fixed Cross: a bit more difficult, because – well, fixed. The fixed signs are resourceful, but they tend to resist change, which makes it harder to grow. Nip's Fixed Cross combined a gregarious Aquarian Sun opposing musical Neptune rising in Leo in square to a Moon-Jupiter conjunction in dark, mysterious Scorpio opposing his MidHeaven in conservative Taurus. Like my father's Cross, Nip's included both luminaries, making it particularly powerful.

The Moon in Scorpio often indicates an intense childhood. I don't think my brother ever really recovered from the shock of being accidentally scalded by boiling water as a child. He learned early in life that the world is a dangerous place. His strong Neptune, conjunct the Ascendant and part of the Fixed Cross, suggests a tendency to escapism. He loved fantasy; like every Washington insider, he drank; but like his father, his drug of choice was work.

All those squares and oppositions of the Cross can be highly motivating, and like all the Weavers, he had a way with words. As a political reporter on the Washington bureau of *The New York Times*, he was known for writing classic *Times* copy, and was so greatly respected by his colleagues that when he died the paper gave him a front-page obituary with a photo.

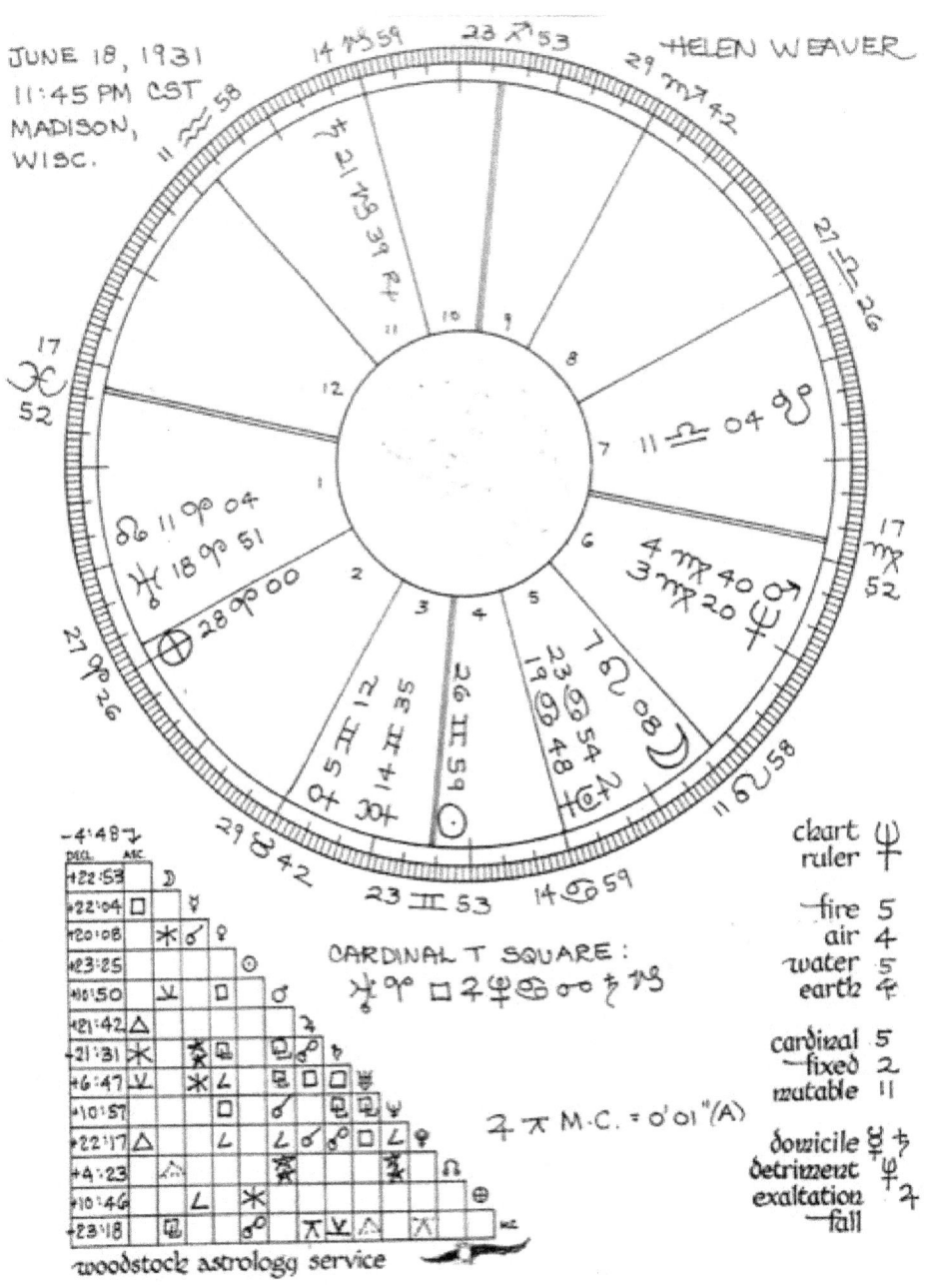

I came along on June 18, 1931, at 11:45 p.m., also in Madison. Mother said I was "on hand by midnight," and the 11:45 time was validated by Charles Jayne, a distinguished American astrologer who specialized in the complicated business of rectification. (Birth times are often inaccurate; rectification is a procedure designed to correct an uncertain time by working backward from important events in the person's life.)

The 11:45 p.m. time gives me the sign of Pisces on the Ascendant. Shy, supersensitive, dreamy, musical, poetic, a bit reclusive, not particularly strong physically, but with a flexible body: I've learned to live with it and to focus on the positive. The Ascendant describes the physical body as well as the personality one shows to the world, and life goes better if one can manage to be comfortable in one's skin.

My Moon in Leo, however, wants attention! I think it was the late Richard Idemon who taught me that when you meet someone whose rising sign is the same as your Moon, there is a kind of shock, for what you secretly long to do is something they can do effortlessly, and in public. As an adolescent, I resented my Leo rising father's tendency to dominate the conversation at the dinner table, and I have always envied both him and my brother for the ease with which they performed in public.

It didn't help, either, that all of my planets except one are below the horizon, giving an introverted cast to the chart. In my chart the only planet above the horizon is serious Saturn in ambitious Capricorn: I don't want to show anything to the world until it is perfect. Unlike most writers, I never submit a book to publishers until it is finished.

But that Saturn is part of a Cardinal T-square, and this major configuration, although it typically spells difficulty at the beginning, has provided me with the motivation to stop feeling sorry for myself and create a life I love. In the T-square, two planets are in opposition and a third forms squares to both ends of the opposition. In my case, Jupiter and Pluto, conjunct in Cancer, oppose Saturn in Capricorn, and this opposition is squared by Uranus in Aries. In effect, I have inherited my father's Cardinal Cross, minus just one leg.

My brother had four conjunctions with my mother's chart and several other harmonious aspects. I have a single conjunction with Mother's chart: my Mercury in Gemini conjuncts her Neptune. Neptune is not a particularly comfortable energy for her, for it squares her Virgo Sun and Mercury. Virgo likes to make distinctions; Neptune likes to dissolve them. But Neptune rules music and poetry and all works of the imagination, and as I learned to play the piano and began reading and writing, Mother and I found common ground. It was Mother who taught me how to practice the piano and who gave me my first copies of *Vanity Fair* and the novels of Jane Austen – and my first diary.

But Mother's and my Sun signs are in square: Gemini finds Virgo too picky and Virgo finds Gemini too scatterbrained and sassy. Mother and I did eventually become good friends, but it took a while.

My brother and I have one lone conjunction between our charts: his Mars in Aries is conjunct my north node. Since his Mars opposes Saturn (a difficult aspect he inherited from his father), I sometimes received the brunt of his irritation, and was dismissed as the hippie black sheep of the family, which only spurred me on to greater acts of rebellion.

Nip (my brother's nickname) and I did share a love of music and poetry, and I see this as the gift of Neptune, which conjuncts his Ascendant and, as the planet that rules Pisces, is the ruling planet of my chart. I still cherish the memory of the summer of 1950, when we sang along to the LP records of *Kiss Me Kate* and *South Pacific* and he read me the poetry of Andrew Marvell.

But with Dad it was a whole different ball game. My stellium of Sun, Venus, and Mercury in Gemini echoes my father's stellium in Gemini, which is all about reading and writing and talking and is interested in everything. My Mercury is conjunct his Neptune within only twelve minutes of arc. Neptune rules mathematics, among other things; Dad taught me math.

With the Gemini emphasis in both of our charts, Dad and I had a lot of Gemini things in common. He was deeply engaged with translation; I became a translator. He translated the complicated ideas of science and mathematics into terms a non-scientist could understand. Like him, I became a writer; and like him, I had a hobby – in my case, astrology – that eventually turned into a profession. In a sense, an astrologer is a translator, someone who renders into understandable terms the foreign language of the stars.

As we have seen, my Sun in Gemini is conjunct Dad's Jupiter, the "greater benefic," and Dad's Sun in Cancer is conjunct *my* Jupiter. As the largest planet in our solar system, Jupiter stands for growth and expansion, in both a material and a spiritual sense. Its qualities are good fortune, optimism, generosity, and humor. It has been assigned rulership over higher education, religion, philosophy, and the law. That double dose of Jupiter between our two charts is a perfect description of our relationship, which was lucky for both of us, and took a philosophical turn, especially toward the end of my father's life.

Actually, the number of Geminis in the extended Weaver family is something of a family joke. A niece, a great niece, two great nephews, a sister-in-law: we not only produce Geminis, we marry them. At a typical Thanksgiving gathering in the old days the proportion of those around the table who had the Sun in that sign was close to 50%. It was never a particularly quiet meal!

In addition to the Gemini connection and the double Jupiter connection I have with my father, there is more. Not only does my T-square mimic Dad's Cardinal Cross, but its planets are in the same degrees of longitude, so that they fit right over the planets of his Cross. We have seen that my Jupiter is conjunct his Sun; but my Jupiter is conjunct Pluto, adding another dimension to the mix. My Saturn is conjunct his Moon and my Uranus is conjunct his Mars.

The Saturn connection reflects the depth and seriousness of the teaching that went both ways between my father and me. The Pluto connection adds a level of intensity that was initially uncomfortable, particularly when I was in my teens and we were locked in a power struggle. And the Uranus connection is all about the excitement of new ideas and the unpredictable and unusual nature of our dialogue about astrology.

My Cardinal T-square differs from Dad's Cardinal Cross in that in my case, none of the bodies involved are personal planets. The outer planets (Jupiter, Saturn, Uranus, Neptune, and Pluto) move so slowly that they are more descriptive of a generation than of any one individual. In fact, that difficult Cardinal T-square of mine is a very accurate expression of the state of the world in the early thirties, with the Great Depression and the rise of fascism in Europe.

According to *Larousse,* the planet at the midpoint of a T-square is the focal point, and represents the means by which the conflicting tendencies are resolved. In my case, that focal planet is Uranus, and since Uranus is in my first house with a strong aspect to the Ascendant, all of this intense and initially conflicting energy does become personal with me.

Uranus in the first house is the signature of the rebel. As soon as I got away from home and out from under the control of my kind but overprotective parents, I went a little wild. I hung out with artists and writers in New York's Greenwich Village, including Jack Kerouac, the father of the group of rebels known as the Beat Generation. I was always attracted to unpopular causes. Perhaps it was inevitable that I eventually found my way to astrology.

When my Uranus opposition arrived in the early 1970s, I left New York City, moved to Woodstock, (capital of the Age of Aquarius – Uranus rules Aquarius), learned to drive, and became an astrologer.

Uranus is strong in my father's chart as well. His Uranus in Scorpio is in close opposition to his Taurus MidHeaven. Since the MidHeaven is the most public place on the chart, it rules a person's aspirations, their career, and what they will be remembered for after their death: their legacy. Taurus is conservative, stable, and nurturing, and accords well with his administrative and philanthropic work at Rockefeller Foundation and elsewhere. But Uranus in Scorpio is all about groundbreaking research. Dad's Uranus in close aspect to his MidHeaven is the signature of the pioneer.

If you look Warren Weaver up in Wikipedia, you will see that after calling him "an American scientist, mathematician, and science administrator," the anonymous author goes on, "He is widely recognized as one of the pioneers of machine translation," paying tribute to that inspired memorandum that was decades ahead of its time. My father also contributed significantly to the development of molecular biology, which he correctly foresaw was the wave of the future.

Uranus rules whatever technology is new at the time. The latest technology sometimes consists of old ideas that have been long neglected but are enjoying a comeback. As the "higher octave" of Mercury, and a very mental planet, Uranus has been assigned rulership over both science and astrology.

Since it takes Uranus approximately eighty-four years to complete its orbit around the sun, Uranus returns to its original position in a natal chart around the age of eighty-four. When

Uranus came within orb of a conjunction to its natal position in my father's chart, he started asking questions about astrology. And in spite of his early training in science with its rational materialist bias, Warren Weaver concluded that this ancient system was worth another look. He had what Jung called "an essential change of attitude."

And as I write this book, Uranus in the sky is exactly where it was when I was born, eighty-four years ago. During the year of my Uranus return I am writing a book about astrology!

Looking back over all of the various connections among the charts of my nuclear family, what have we learned about the way astrology works with heredity?

In the first place, the Gauquelin effect involving the rising or culminating of certain planets with a frequency greater than chance gets a vote of no confidence from the Weavers. Even were we to expand his parameters to include any and all planets in angular houses (first, fourth, seventh, and tenth) we would still see no evidence of what he called planetary heredity.

My father has one angular planet: Uranus in the fourth house.

We cannot apply the Gauquelin effect to my mother, since we have no idea what her angles were.

My brother has four angular planets: Neptune rising, Moon and Jupiter conjunct in the fourth house, and the Sun in the seventh, all part of his fixed cross. If we were to expand the Gauquelin parameters to include planets near all of the angles (not just rising and culminating, but conjunct the fourth and seventh house cusps as well) then his Moon-Jupiter conjunction would agree with his findings: the Moon prominent in the charts of writers, and Jupiter in the charts of actors. My brother was both. But that's stretching a point.

I have one angular "planet": the Sun in the fourth house.

I guess you could say that the Weavers tend to have planets near the fourth house cusp. The fourth house cusp – or the I.C., *Imum Coeli*, "bottom of the sky" as it is traditionally known – is exactly opposite the M.C., or MidHeaven, which rules career. Here, we can find common ground, for my father, my brother, and I were/are all ambitious, wanting to make a mark or make a difference, and willing to work hard to give something back to the world.

I doubt if Gauquelin would be impressed.

But there are other ways to look at the possibility of heredity in family charts. Larousse mentions conjunctions and oppositions, especially those involving Sun, Moon and Ascendant, and major configurations that reappear in the charts of family members.

My father and my mother have three conjunctions that we know about (not counting the Moon's nodes or the modern planets they share with all members of their graduating class): Ascendant-Venus; Saturn-Saturn; MidHeaven-Mars; plus the possibility of aspects to her angles and/or major configurations that might show up if we had her correct time of birth.

My father and my brother have six conjunctions between their charts: Ascendant-Ascendant; Ascendant-Neptune; Saturn-Saturn; Uranus-Moon; Mars-Mars; MidHeaven-MidHeaven. In addition, both charts have Crosses in which both Sun and Moon are involved. They both have Mars opposite Saturn, in the same houses, the same signs, and close to the same degrees.

My father and I have six conjunctions between our charts: Mars-Node; Mars-Uranus; Neptune-Mercury; Jupiter-Sun; Sun-Jupiter; and Moon-Saturn. In addition, I inherit his Cardinal Cross minus one leg, and we share a stellium in Gemini.

If we turn to my mother and brother, in spite of the lack of information we have for her, we see that they have four known conjunctions between their charts: Saturn-Saturn; Uranus-Moon; Mars-MidHeaven; and Venus-Ascendant – and who knows what else.

But Mother and I have only one conjunction, the Neptune-Mercury conjunction which I would have with all members of her graduating class. And my brother and I have only two conjunctions: his Mars-my North Node, and his South Node-my Ascendant; and both of these are connections we would each have with all other members of the other's graduating class. In other words, these two connections are impersonal. Although he and I were only eight years apart, it was as if we belonged to different generations: his, the Great Generation, and mine, the Beat Generation and their heirs, the hippies.

It seems to me that this is an accurate reflection of the family dynamic. I don't expect it would convince a skeptic; but this is the sort of thing that anyone who studies astrology comes across over and over again. Sooner or later, they begin to suspect that there's something to it.

APPENDIX NO. 2
A HISTORY OF ASTROLOGY

Anything that endures for hundreds or thousands of years (at least in part by voluntary enjoyment rather than forced study) must contain something of value.

Stephen Jay Gould

Dad thought that our book should include a chapter on the history of astrology.

Indeed, it was the very fact that astrology has had such a long history – the persistence of belief by thousands of people down through the ages – that had impressed him long before he came to accept the idea that the astrological hypothesis – what he called "Assumption B" – was worthy of consideration by science.

I was never terribly enthusiastic about the prospect of writing this chapter, partly because I felt inadequate to the task, and partly because I assumed that there were so many good histories of astrology readily available.

In the latter case I was mistaken; Most mainstream academic historians are so hostile to the subject of astrology that they don't bother to study it, and consequently their work suffers from errors and omissions and often from a snide attitude reflecting their materialist bias. And when astrologers take up their pens to write the history of their practice, they have their own bias to contend with. Both versions of the history of astrology have been hampered by substantial lacunae in the written records.

Happily, all this is starting to change. In 1992 Kepler College was incorporated by a group of astrologers and academics who believed that astrology deserved a place in academia. While it cannot be said that astrology has been reinstated in the universities from which it was banished in the seventeenth century, nevertheless a certain cross-fertilization has begun. In 1993 a group of astrologers including Robert Hand founded Project Hindsight, a publishing company whose mission is to translate ancient texts from Greek, Latin, Hebrew, Arabic, and Sanskrit into modern English. And then in 1997 the American astrologer Robert Hand founded the Archive for the Retrieval of Historical Astrological Texts (ARHAT), a formal archive, library and publishing company specializing in the history and science of astrology. Combined with some recent work by astrologers who are also academics, these projects have all the earmarks of a renaissance. At long last, astrology is being carefully studied by responsible scholars who are no longer hostile to the subject.

In short, I have come across some exciting recent literature on astrology's past that avoids the usual pitfalls and has reminded me of Dad's wish and motivated me to honor it. What I would like to offer is a brief history of Western astrology, with special emphasis on two periods in which that history has taken a dramatic turn: the seventeenth and eighteenth centuries, when astrology became divorced from astronomy and lost favor with the educated classes, and our own time, when there are some very tentative beginnings of a possible reintegration of the two sciences of the stars.

I mean to focus on Western astrology, not because I consider it superior to its Eastern counterpart – far from it – but for two entirely different reasons. In the first place, because my forty-odd years of study and experience have been in Western astrology; but more importantly, because it is Western astrology that has suffered the unfortunate schism that is the *raison d'être* of this book. It is my understanding that in India, for example, astrology has enjoyed an unbroken tradition of popularity and respect. In India today, both astronomy and astrology are referred to as *jyotish*: "the science of light."

Since I'm not an expert on this subject, I'm going to rely heavily on the work of those who are, especially Robert Hand in America and Nicholas Campion in England.

1) Origins

Astrology is at least five thousand years old. Since it predates recorded history, its origins are lost in the mists of time, but it is generally believed that astrology as we know it in the West first developed in Mesopotamia, the land between the Tigris and Euphrates rivers (*mesopotamia* is Greek for "between the rivers") commonly known as "the fertile crescent." This area is often referred to as "the cradle of civilization," since it is the first known site of settled agricultural communities. The Mesopotamians are believed to have been the first people to develop writing, mathematics, and astronomy. The earliest form of writing was cuneiform, in which wedge-shaped characters were pressed onto clay tablets. The first calendars were made of stone and the first horoscopes were written in clay.

The first astronomer/astrologers were priests who studied the movements of the Sun, Moon, stars, and planets in order to predict the weather and anticipate the seasons – and to determine the will of the gods who were identified with the planets. According to their theology, conditions in the heavens corresponded to conditions on Earth: "As above, so below."

Of course, the Mesopotamians were not the only ancient peoples who looked to the sky for omens and began to connect the motions of the heavenly bodies with events on Earth. As Robert Hand has pointed out:

... some kind of astrology is nearly universal among ancient peoples and is not limited to either one time or place as its point of origin. Almost every ancient people had some system of examining the heavens for divinatory purposes: Native Americans, Greeks (long before they encountered Mesopotamian astrology), the peoples of India, whoever it was that built Stonehenge and New Grange in the British Isles, and the ancient Nordic peoples, to give a partial list.[63]

But it is Hand's thesis that "astrology as we know it came into being only once in time and in one place; the place is Mesopotamia (roughly modern Iraq). . .." By "astrology as we know it" Hand means "horoscopic astrology": astrology based on a chart cast for a given time and place.

There are four main branches of horoscopic astrology that have survived the centuries and are with us today: mundane astrology, formerly known as "revolutions," probably the oldest branch, which studies planetary cycles as they relate to world events; natal or genethliacal astrology, which studies the birth charts of individuals to gain insight into their character and destiny; electional astrology, concerned with choosing the most favorable time to schedule an event or launch an enterprise; and horary astrology, formerly known as "interrogations," a divinatory technique in which questions are answered by casting a chart for the time they were asked.

British astrologer Nicholas Campion, in his authoritative two-volume *History of Western Astrology,* makes an important point about the common assumption that these early stargazers who saw intimate connections between the cosmos and the human psyche were the victims of primitive thinking:

The whole notion of "primitive" as a moral or intellectual condition needs to be well and truly abandoned. . .. For the pre-modern mind . . . the world is conceived as an integrated set of living beings, in which there is no difference between the psychic and the physical.[64]

So it was not just the need to predict weather for agricultural purposes or anticipate the will of the gods that explains the birth of astrology, but the broader world view of these people:

While arguments that early humans observed the stars because they either needed to herd their animals, plant their crops (in the Neolithic period), placate natural forces or worship divine powers may have some truth, they are all functional and reductionist, seeking single, useful explanations whilst ignoring the bigger picture.[65]

For these people,

the entire cosmos was thought to be a single living entity, all of whose constituent parts, from the psychic to the physical, and from gods to planets, people, plants, and minerals, were interdependent and interlinked.[66]

The universe was alive and everything in it was connected. The bigger picture is the world view that has been dismissed as "magical" but which might more accurately be described as ecological.

Hand writes that in the beginning, Mesopotamian astrology

was much like that of other cultures, a simple examination of the heavens for omens that might affect the kingdom. . .. What made the Mesopotamians different is that they began at an early time to make systematic observations of phenomena with an eye to finding regular patterns in the heavens that might correlate with patterns in human events.[67]

Dating from around 750 BCE, and for a period of over 700 years, the Mesopotamian astrologers kept diaries, a sort of log of their regular observations of the heavens. They (especially the Babylonians) eventually got to the point where, as Hand puts it,

> . . . based on observed recurrent cycles of the planets, they could reasonably accurately estimate the positions of the planets at any time in the future.[68]

Such observations were a necessary but not a sufficient basis for astrology. In order to have astrology, you have to have a zodiac. Hand writes that in the beginning,

> . . . they simply recorded planets as being so many degrees from a star. . . . This is *de facto* a sidereal observation, but it is not a zodiac! A zodiac requires a fiducial point: a point on the circle from which measurements are made.[69]

For the tropical zodiac this point would be zero degrees of Aries, or the position of the Sun at the vernal equinox. And for most zodiacs, the 360 degrees of the ecliptic – the Sun's apparent path around the Earth – is divided up into workable sections, e.g., twelve 30-degree signs in the West or twenty-seven lunar mansions in the East.

According to the Dutch mathematician Bartel Leendert van der Waerden (1903-1996), author of a two-volume history of mathematics entitled *Science Awakening*, the evolution of astrology went through three phases. Hand summarizes them as follows:

> The first phase consists of the omen lore that we have already described.
> The second phase is closely related to this, but has a zodiac in the modern sense of twelve 30-degree signs. There is no personal horoscopy in this middle level, but great attention is paid to the transits of Jupiter through the signs at the rate of approximately one sign per year. . . . There are, of course, no houses of any kind. Van der Waerden dates this middle phase as being from about 630 to 450 BCE. The zodiac at this point is clearly a sidereal one [i.e., based on the stars rather than on the spring equinox, the point where the ecliptic intersects the equator]. . ..
> The third phase is horoscopic astrology. Various ancient sources mention "Chaldeans" [i.e., Babylonians; the term "Chaldean" came to be synonymous with "astrologer"] who cast birth charts for various persons, including Diogenes Laertius, who said that according to

Aristotle, a Chaldean forecast Socrates' death from his birth chart, and that Euripides' father also had his son's chart read, getting a forecast of his brilliant career.[70]

According to Campion, the earliest evidence for the existence of the zodiac is a tablet dated to 475 BCE, while a tablet dated to 419 BCE gives a list of planets in the signs: in other words, the first known ephemeris.[71] Campion believes that the signs of the zodiac evolved as a system of measurement, but eventually came to have meanings. By the fifth century BCE the planets seem to have acquired the personalities that they were to have for the Greeks, and which have endured to this day: Venus is kind, Jupiter is generous, Mars is hot-tempered, Saturn is a stern teacher.

The oldest known birth chart, written in cuneiform on a clay tablet, has been dated April 29, 410 BCE. Cast for "the son of Shuna-usur, one of Shuma-iddina, descendant of Deke," it gives the following planetary positions:

. . . moon below the horn of the Scorpion (that is, Libra), Jupiter in Pisces, Venus in Taurus, Saturn in Cancer and Mars in Gemini. Mercury was invisible, being too close to the sun to be seen, and so was ignored. The sun itself was not mentioned.[72]

The next known birth chart, which is dated 263 BCE:

. . . was the first to include planetary positions recorded by degree, indicating an increasing concern with mathematical precision. It also includes the sun as well as Venus, which was close to the sun and hence invisible, indicating that the astrologer may have been working from tables of positions computed in advance, rather than direct observation.[73]

This third phase lasted a long time, with increasing precision in the positions of the planets.

Following the early Babylonian period, there is a considerable gap in the written records available to historians; but according to Robert Hand, who has unearthed and translated, or seen to the translation of, some previously neglected Hellenistic (i.e., Greek speaking) texts, the internal evidence of this newly discovered material suggests that the true birthplace of horoscopic astrology as we know it may have been Egypt.

While it is well documented that as early as the fourth millennium BCE the Egyptians had a thorough knowledge of astronomy, very little is known about the origins of astrology in Egypt. Apparently the priests who had a monopoly on knowledge of the universe were either very secretive about it or preferred oral to written transmission, or both.

There are hints to the antiquity of the Egyptian tradition. For example, the Great Sphinx at Giza which dates from the third millennium BCE embodies the four fixed signs of the zodiac: It has the head of a human being (Aquarius), the body of a bull (Taurus), the paws of a lion (Leo), and the wings of an eagle (Scorpio).

When the Chaldeans moved into Egypt after the Persian conquest (525 BCE), Mesopotamian influence dramatically increased. While it is not known exactly what the Egyptians added to Babylonian astrology, Hand states that internal evidence of the Hellenistic texts suggests that the use of a rising degree (i.e., an Ascendant) may have predated the Greeks; and the Hellenistic writers attributed the use of houses to Hermes, which indicates Egyptian origins. Hermes Trismegistus (Hermes "thrice great"), was an archetypal figure who represented a merging of the Greek god Hermes and the Egyptian god Thoth, and was the alleged author of the Hermetic writings, a group of sacred texts of a Gnostic kind.[74]

Hand believes that the aspects are probably also Egyptian, and that the lots (the so-called "Arabian parts," a group of mathematically derived points which have traditionally been attributed to the Arabs) are almost certainly Egyptian, as well as most systems of rulership:

At any rate it is quite likely that the entire apparatus of horoscopic astrology was in place by 1 CE, and quite possibly several centuries earlier.... Whatever may have been the language of Egyptian astrology when it first began to come into being, by 1 CE it was Greek.[75]

Before we leave the Egyptians, it is worth remembering that what they added to Mesopotamian astrology went beyond the technical to embrace the metaphysical. Campion lists the few surviving texts that provide clues to Egyptian religious beliefs – the inscriptions on the walls of the tombs inside the pyramids; the funerary spells inscribed on coffin lids, papyri, and mummies;

the set of texts known as *The Book of the Dead*; and the astronomical ceilings portraying the heavens – and concludes:

Collectively, these texts establish for the first time in written form the belief, which may be many thousands of years old, and which has certainly lasted until the present day, that one's spiritual destiny is connected to the stars. And this is critical for our understanding of Western astrology; while the Babylonians devised a technical system for reading the sky and interpreting divine intentions, the Egyptians provided a meaning and purpose in which the rationale was one's own individual connection with the stars.[76]

2) The Greeks and the Romans

The tendency of modern scholars is to idealize the "rational" Greeks and to believe that they were uncorrupted by astrology until the conquests of Alexander the Great (334-23 BCE) opened the Hellenistic world to Chaldean influence. As Campion points out, the Greeks are admired as the source of all that is rational and scientific about the ancient world while the Orientals are blamed for all that is irrational and superstitious – especially astrology.

Typical of this mindset is the British scholar Gilbert Murray, who wrote in 1935: "astrology fell upon the Hellenistic mind as a new disease falls upon some remote island people."[77]

This is a distorted reading of the evidence. In the first place, it is undeniable that many of the underlying ideas of traditional Western astrology are the legacy of classical Greek philosophy. As early as the sixth century BCE, the great mathematician and philosopher Pythagoras (585-ca. 495 BCE), who made a profound study of the relationship between mathematics and music, gave the world not only the Pythagorean theorem but also the semi-mystical doctrine of the Music of the Spheres. Pythagoras taught that the spaces between the orbits of the planets corresponded to the frequencies of the tones of the musical scale, and that as the planets moved along their orbits, which were conceived as concentric crystalline spheres, their motion created a kind of ethereal harmony. The idea of the Music of the Spheres has captured the Western imagination down through the ages.

Pythagoras also taught that each human being was a microcosm: a miniature universe reflecting the macrocosm, the universe as a whole. An important idea that underlies the astrological world view, the notion of man as a microcosm was echoed by Plato in his *Timaeus* and was later embodied into Hermetic theory, a body of correspondences that engaged the medieval mind and survives in the esoteric astrology of today.[78]

The Greek philosopher, poet, and scientist Empedocles (ca. 495-435 BCE) taught that the basic building blocks of matter were the four elements of earth, air, fire, and water, an idea that became inseparable from the astrological world view.

Plato (ca. 428-348 BCE), the towering figure of Greek philosophy, had a complex attitude toward astrology. He thought that astrology could be useful for the management of the ideal state, but he was highly critical of the detailed predictions of the Babylonians. Campion calls Plato "the first in a long line of Western astrological reformers, of whom the last was Johannes Kepler in the seventeenth century."[79]

Plato was essentially a theorist: he distrusted all knowledge that was obtained by observation. Indeed, he believed that it was really impossible to know anything for sure. He was perhaps the first great skeptic in recorded history.

At the same time, Plato believed in the unity of the universe, which is the basic assumption of astrology. The Platonic cosmos was a single, unified, living, and divine entity in which all the parts were interconnected. Plato accepted the Pythagorean doctrine of the Music of the Spheres. He believed that the planets moved in perfect circular orbits around a motionless central Earth.

In Plato's universe change and motion represented imperfection, whereas permanence was divine. The fixed stars were unchangeable, and therefore closest to God; the Sun and the planets, the stars that "wandered"(from the Greek *planasthai*, to wander), were at various degrees of separation from God; the Moon, whose rapid changes were most apparent, was even further from God; and on the "sublunary" ("beneath the Moon") Earth, in the center of the system, in spite of the Earth's apparent stillness, all was change and decay.

In spite of his skepticism about the possibility of knowing anything, let alone predicting the future, Plato believed that the soul came from God by way of the stars and that the planets could thus provide a pathway back to God. This "spiritual cosmology," as Campion puts it,

> ... established a model for the relationship between the soul and the stars which ... provided a rationale for astrology [and] which is still evident in the number of New Age astrology books which include the words "spirit" and "soul" in their titles.[80]

There were Greeks who proposed that it was the Earth that moved rather than the Sun, notably, Aristarchus of Samos (ca. 310-250 BCE), who argued that the Earth rotated around the Sun. But Plato's fixed Earth cosmology was adopted by his student, Aristotle, and by Ptolemy, the Alexandrian astronomer and astrologer who codified the astrology of his day, and it was this cosmology that prevailed throughout the Middle Ages and into the Renaissance.

The Greeks imported astrology from the Babylonians between the fifth and third centuries BCE. In the early third century BCE, the Babylonian priest Berosus set up a school of astrology on the Aegean island of Kos.[81] It was also around this time that work was begun on the massive Library of Alexandria under the Greek pharaoh Ptolemy 1. From the second century BCE on, most astrological texts were written in Greek, which had become the *lingua franca* of the Hellenistic age.

Whereas Babylonian astrology had been predictive and deterministic, the Stoics, a school of philosophers founded in Athens in the third century BCE, taught that "the wise man rules his stars," an idea that is found in many modern texts. These two threads – fatalism and free will – persisted throughout the Middle Ages and underlay the ambivalent attitude of the Christian church toward astrology.

The Roman philosopher and politician Cicero (106-43 BCE) was highly critical of astrology. His book *De divinatione,* which dealt with all forms of divination, including astrology, was in the form of a dialogue between himself and his brother Quintus. Quintus argued in favor of divination while Cicero took the skeptical position. Cicero's arguments against astrology formed the first full-scale attack on the subject in recorded history. They have been used by critics of astrology down to the present day, including the authors of the *Humanist* manifesto.

To give only two examples of these classic arguments, Cicero raised the twins problem:

if two babies were born at the same time and in the same place, what accounts for the differences in their characters and destinies?[82] Cicero also cited the impossibility of action at a distance: the planets are too far away to have an effect on matters here on Earth.

Cicero's critique – although it did appeal to a certain element in the Christian church who regarded astrology as a tool of the devil – failed to undermine its popularity in an age that was imbued with the idea that each human being was a microcosm of the universe and that the pathway to God was through the stars.

3) The Christian Era

By the dawn of the Christian era most of the signs of the zodiac had acquired the meanings that would be familiar to a modern astrologer. And two other important factors had been added to the practice of astrology in the Hellenistic age: the *horoscopos*, the degree of the zodiac rising at the Eastern horizon at the moment for which the chart is cast (in modern language, the Ascendant), and the places, or houses, the divisions of the wheel into twelve sections according to one of several house systems. Over the years, the word "horoscope" came to stand for the whole map of the heavens at a given point in time and space.[83]

The next towering figure in the history of Western astrology is Claudius Ptolemy of Alexandria, born around 70 CE. Although he was an avid astrologer, he may have contributed to astrology's eventual separation from science in several ways.

In the first place, he divided astronomy from astrology by treating what was hitherto one and the same subject in two different books. His *Mathematike Syntaxis*, better known by its laudatory Arab title *Almagest* ("the greatest,"), a treatise on the astronomical knowledge of his day, was all about calculating the motions of the stars and planets, whereas his *Tetrabiblos*, a treatise on the astrological lore of his day, was his attempt to compile and organize the vast body of existing material into a unified whole. The first book was concerned with measurement, the second with meaning.

In the second place, Ptolemy is credited with (or blamed for) eventually undermining the credibility of astrology by adopting the tropical zodiac, which is based on the vernal equinox and the seasons, rather than the sidereal zodiac, which is based on the constellations and takes into account the precession of the equinoxes.

Larousse defines precession of the equinoxes as:

The continuous shift of the equinoxes backward through the sidereal zodiac, as a result of the slow revolution of the Earth's axis of rotation about the ecliptic pole, itself caused by the gravitational attraction of the Sun and Moon on the Earth's equatorial bulge.[84]

For example, according to sidereal astrology my mother, who was born with her Sun in the early degrees of Virgo in the tropical zodiac, would actually have the Sun in late Leo.

Credit for discovering the phenomenon of precession has been given to the great Greek astronomer Hipparchus of Rhodes of the second century BCE, but it may have been known long before that. It has been a favorite weapon of critics of astrology, who maintain that the signs of the tropical zodiac no longer express the "real" positions of the planets, and who assume that astrologers are ignorant of basic astronomy. In fact, astrologers are well aware of precession – indeed, there is a respectable school of sidereal astrology here in the West – but those who favor the tropical zodiac believe that the constellations are primarily a visual construct and have no natural integrity as a frame of reference.

Finally, Ptolemy's system perpetuates the Earth-centered cosmology of Plato and Aristotle, and this has also helped to bring about the schism between Western astrology and science.

An important figure in the early Christian era was the Greek Neoplatonist philosopher Plotinus (203-270 CE). The Neoplatonist school was inspired by the ideas of Pythagoras, Plato, and Aristotle, as well as the esoteric teachings attributed to the legendary Hermes Trismegistus.[85] In his tract *On Whether the Stars are Causes*, Plotinus made a subtle distinction between an astrology of causation, in which the planets cause things to happen, and an astrology of signs, in which the planets accompany events as signs.

Plotinus believed that the influence of the stars could be mitigated or even overcome by those who were spiritually evolved. Again, we encounter the Stoics' idea that "the wise man rules his stars," which persists down to the present.

When we come to the early history of Christianity the role of astrology becomes extremely complex. The Bible itself is nothing if not ambivalent on the subject. Several passages in the Old Testament condemn astrologers along with soothsayers and magicians:

There shall not be found among you any one . . . that useth divination, or an observer of times, or an enchanter, or a witch. . . . For all that do these things are an abomination unto the Lord. . . .[86]

Leviticus, Isaiah, and Jeremiah are equally hostile to astrology.[87] Yet in the very first chapter of Genesis, the first book of the Bible, God says:

Let there be lights in the firmament of the heaven to divide the day from the night; and let them be for signs, and for seasons, and for days, and years. . . .[88]

There is some reason to believe that the prophet Ezekiel (the one who "saw the wheel way up in the middle of the air") was himself an astrologer. In his first vision, God's chariot was drawn by a man, a lion, an ox, and an eagle – the symbols for the four fixed signs of the zodiac.

And in the gospel according to Saint Matthew in the New Testament, the birth of Christ is heralded by the appearance of the Star of Bethlehem, and the three wise men who follow the star are *magi*, from the Greek word for astrologers. How could astrology be an abomination when the birth of the Savior had been announced by a star?

Many modern astrologers suspect that the "star" of Bethlehem may have been a conjunction of Jupiter and Saturn in Pisces around 6 or 7 BCE, an event that would have been seen as an important celestial omen. The New Testament has nothing negative to say about astrology, and indeed, Luke declares that when the temple of Jerusalem is destroyed by the Gentiles, "there shall be signs in the sun, and in the moon, and in the stars."[89]

But the early Christian writers from Paul forward believed that salvation was meaningless without free will, which astrology seemed to deny. The church fathers who developed Christian doctrine under Roman rule condemned astrology along with all forms of pagan superstition. The church's position remained unchanged down to the twentieth century and is reflected in the Roman Catholic catechism of 1994:

All forms of divination are to be rejected: recourse to Satan or demons, conjuring up the dead or other practices falsely supposed to "unveil" the future. Consulting horoscopes, astrology, palm reading, interpretation of omens and lots, the phenomenon of clairvoyance, and recourse to mediums all conceal a desire for power over time, history, and, in the last analysis, other human beings, as well as a wish to conciliate hidden powers. They contradict the honor, respect, and loving fear that we owe to God alone.[90]

On the persistence of the church's objections to astrology Campion comments:

It is impossible to deal adequately with medieval and Renaissance astrology without considering the collision between astrology, which suggested that the individual has a direct relationship with the cosmos, and Christianity, which insisted that this must be mediated via the Church.[91]

The first explicit condemnation of astrology by the Christian church was issued by the Council of Laodicea (364 or 367), which decreed that priests could not practice astrology. The Council of Toledo in 400 threatened anyone who believed in astrology with excommunication. Yet through the ages, many popes consulted astrologers to learn when they or the cardinals would die.

Saint Augustine (354-430), the great convert to and advocate for the Catholic faith, is an interesting case. As a young man, Augustine belonged to the Manicheans, a Gnostic sect whose writings were full of astrology. But after his conversion, he decided that the Manicheans were guilty of a kind of pantheism:

... [They] came too close to worshiping the sun and the moon, in other words, worshiping the creation rather than the creator, a criticism of astrology which is repeated in evangelical Christian literature to this day.[92]

In his *Confessions* (ca. 397), Augustine denounced astrology because it undermined the authority of God and seemed to absolve humans from moral responsibility for their actions. In *The City of God* (410), he denounced it again, this time because it undermined the authority of the Roman Emperor, who was now regarded as God's instrument on Earth – an idea that persisted until the Protestant Reformation in the sixteenth century.

Augustine thought that astronomy was a waste of time, since it did not help people to attain salvation, which was the goal of religion. And yet the sacred calendar of the Christian church was based on the astronomy it inherited from the pagans. By 336 CE, the birth of Christ was celebrated at the Winter Solstice and his resurrection at the Vernal Equinox, as a result of the church's coopting of the rituals of the Roman religion of Sol Invictus ("the Unconquered Sun").

4) The Middle Ages

With the fall of the Roman empire, and from the fifth century on, astrology suffered a decline in Christian Europe. Campion attributes this decline partly to religious intolerance but mainly to the decline in literacy – especially the knowledge of Greek – in the cultural turmoil following the barbarian invasions. The Greek-speaking Eastern empire escaped Germanic domination, but the Latin-speaking West was overrun by barbarians. Only fragments of classical philosophical texts were available. All the great works of horoscopic astrology were lost to the early medieval world.

It is no longer acceptable to talk about the "Dark Ages," but when I took a course in European history at Oberlin back in the fifties I remember learning that it was the Christian monasteries, with their libraries of ancient texts and their monks copying them out by hand, that kept learning alive during that " benighted" era.

There must be some truth to this, but according to more recent scholarship the true heroes of this period, at least in terms of astrology, were the Arabs. From the eighth century on the Islamic world was the center of astrological activity. The Arabs translated Greek texts into Arabic, synthesized the teachings of the Hellenistic astrologers, and added technical discoveries of their own, many of great mathematical sophistication. The Arabs were the first to make exact calculations of the positions of the planets, and their science confirmed the precession of the equinoxes.

The Arabs have been credited with the invention of the Parts, or Lots, mathematically derived points on the wheel of the horoscope which are sensitive even if no planet is there. For example, the Part of Fortune represents a mathematical relationship between Sun, Moon, and Ascendant and is considered a very fortunate degree, a point of integration in the chart.

Hand believes that the Parts are almost certainly Egyptian in origin and that other elaborate techniques, such as Translation of Light,[93] may have been Persian, but it was the Arabs who were responsible for developing and recording these techniques so that ultimately they could be discovered and adopted by astrologers in the West.

Credit for the survival of astrology in the early medieval period must also be given to the Hebrews who, despite the disapproval of the Old Testament prophets, had a tradition very similar to that of the Arabs. Modern astrologers are especially indebted to the Kabbalists of medieval Spain, whose sacred numerological teachings helped to preserve the mysticism of the classical world. The Kabbalah remains a vital source of the modern esoteric tradition that is sometimes referred to as New Age.

Finally, Western astrology owes a great debt to India, whose traditions preserve much that would otherwise be lost.

While astrology was flourishing in the Islamic world, especially at the Persian court in the magnificent city of Baghdad, in the Frankish kingdom of Charlemagne (ca. 742-814), the first of the Holy Roman Emperors, astrology enjoyed a revival of sorts, along with mathematics, astronomy, and classical learning. Charlemagne himself studied astrology with Alcuin, the greatest scholar of his age.

But the kind of astrology that was known in the West at that time had nothing to do with horoscopes or the soul. The precise calculation of the positions of the planets had been lost, so the casting of horoscopes was not possible. Only *computus* – the calculations that were necessary to manage the liturgical calendar by observation of the sun and moon – had survived, along with a simplified form of electional and horary astrology.[94]

In the ninth century interest in astrology increased, and after the eleventh century Arab astrology flooded into the Western world. Around 1030 the introduction of the astrolabe[95] into Western Europe permitted more precise measurement of the motions of the planets. The first ephemerides and tables of houses were produced in eleventh-century Spain by Islamic scholars. Early Latin astrological texts survived in the libraries of the cities of Islamic Spain as well as in the monastic libraries of the Christian world.

In the eleventh century Catholic scholars translated Greek and Hebrew texts, including the Kabbalah, into Latin. The secret teachings of the Kabbalah revealed a world of correspondences and mystic meaning. Each of the twenty-two letters of the Hebrew alphabet corresponded to a number, and the numbers in turn corresponded to the signs of the zodiac. God had engraved the constellations on the vault of heaven as signs. The world view of the Kabbalah accorded with the spiritual living cosmos of Plato.

The early eleventh century *Liber Planetis et Mundi Climatibus* (Book of the Planets and Regions of the World) was the first astrological text published in Europe. In 1126 Arab astronomical tables and astrological texts were translated into Latin. In 1138 Ptolemy's *Tetrabiblos* was translated into Latin, and later in the twelfth century a translation of Ptolemy's *Almagest* appeared. Hermetic texts were translated into Latin as well.

In the twelfth century we begin to find images of the human body with the signs of the zodiac imposed on the various parts, from Aries on the head to Pisces on the feet. A knowledge of astrology was essential to medicine, for both diagnosis and treatment.

By the twelfth century astrology had become part of the curriculum of the cathedral schools and universities, and the four main branches of horoscopic astrology that have survived into modern times – electional, mundane, horary, and natal – were being practiced. Most people didn't know their date of birth, much less the time of day they were born, so interrogations, in which a horary chart was cast for the time the question was asked, were widely used to answer questions.

The twelfth century saw a renewed interest in classical knowledge. The translation of the complete works of Aristotle took place in the twelfth and thirteenth centuries.

The famed Jewish philosopher and physician Moses Maimonides (1135-1204) believed in a general relationship between the motions of the planets and events on Earth, but was violently opposed to the casting of horoscopes, a view that had an impact on Thomas Aquinas, the great Christian philosopher of the next generation. Maimonides blamed the astrologers, with their implicit undermining of God's power, for the destruction of the Jewish state and for the diaspora.

Roger Bacon (1214-1294), the English philosopher and Franciscan friar who was an early champion of the scientific method, was a practicing astrologer. He developed the theory of elections, the branch of astrology concerned with finding the best time to begin an undertaking. He borrowed from the Arabs the idea that the rise and fall of historical epochs corresponded to the Jupiter-Saturn cycle. Bacon got in trouble with the Church for his pro-astrology views.

The first major astrological text of Medieval Europe was thirteenth-century astrologer Guido Bonatti's *Liber Astronomiae*. Bonatti (1223-1300), the most eminent astrologer of his time, believed that astrology could only provide part of the solution to life's problems, and that astrologers should pray to God for guidance.

By the thirteenth century astrology was practiced on all levels of society. Intellectuals of the Middle Ages took the validity of astrology for granted. They still argued about the degree of determinism it involved, and many found the solution to the dilemma in the idea that the "rational soul" can rise above the influence of the stars.

The two great thirteenth-century Dominican theologians and scholars, Albertus Magnus (1193-1280) and Thomas Aquinas (1225-1274), were both pro-astrology. Albertus Magnus, who advocated the peaceful coexistence of science and religion, subscribed to the very modern idea that the stars impel, but do not compel, thus saving free will in a kind of Augustinian compromise.

Aquinas, who laid the foundations of Catholic philosophy until the seventeenth century, persuaded his colleagues that astrology was acceptable. In his *Summa Theologica* he held that the stars exerted an influence on the human body but not the soul, so that it was possible to rise above it. Thus he allowed room for free will and moral choice.

The ambivalent attitude of the Church toward astrology is nowhere better exemplified than in the great Italian poet Dante Alighieri (1265-1321). In his *Divine Comedy* Dante confined the astrologers (including Guido Bonatti) to hell with their heads turned backwards as punishment for looking into the future, which only God could do. But Dante believed in astrology and boasted that his own genius was the gift of his Sun in Gemini. To Castor and Pollux, the two stars that make up the constellation of Gemini (the Twins), he confessed, "To you I owe such genius as doth in me lie."[96]

Nicholas Campion resolves this apparent contradiction thus:

[Dante] . . . believed in astrology in the looser sense that the cosmos contained meaning and significance, but not in the meticulous certainties that came with the art of horoscope interpretation. His problem with Bonatti was not that the astrologer predicted the future from the stars but that he did so with such precision.[97]

Interestingly, in *The Divine Comedy*, the sun is the center of the cosmos, which Campion calls "a form of pre-Copernican spiritual Copernicanism."

Astrologers not only predicted the future, but helped their clients to alter it. Here we get into the realm of magic, whose association with astrology was at least partially responsible for astrology's eventual downfall. The twelfth and thirteenth centuries saw a revival of the Neoplatonist doctrine of sympathies. If everything in the universe is connected by a web of sympathies, including words, then incantations, as well as talismans created at the right time, have power and can be used for magical purposes. The key text for this sort of thing was an Arab work translated into Latin in 1256 as the *Picatrix*.

From the beginning of the fourteenth century astrology was an important part of the arts and science curriculum at the great medieval and Renaissance universities, including Padua, Bologna, and Paris. The terms *astronomia* and *astrologia* were often used interchangeably to refer to both sciences of the stars.[98]

The work of Geoffrey Chaucer (ca. 1343-1400), considered the father of English literature, is full of astrology. His many references to it in his magnum opus *The Canterbury Tales* show how popular it was, how much a part of the culture. The plot of the Merchant's Tale hinges on astrology. The Wife of Bath's tale shows a rather detailed knowledge of astrology (e.g., her lecherous nature is ascribed to Mars in Taurus) but not of astronomy (her horoscope has Venus and Mercury in opposite signs, an impossibility). The first use in English of the technical astrological terms "ascension" and "declination" occurs in *The Canterbury Tales*.[99]

The kings of England and France were patrons of astrology; some were astrologers themselves. The Holy Roman Emperor Wenceslaus (1378-1400) of the Christmas carol used astrology.

5) The Renaissance

In the fifteenth century, astrology was an important part of the education of physicians. They needed to learn astrology in order to calculate charts for the onset of an illness. A "decumbiture" (literally, "lying down") horoscope was cast for the moment the patient took to her bed. The physician would also study the natal chart of the client to determine what herbs were indicated and would schedule surgery, bloodletting, or leeches according to the phases of the moon.

In 1455 Johann Gutenberg (ca. 1398-1468) produced the first European book to be printed with movable type, the so-called Gutenberg Bible. The invention of printing was an historic event that played a major role in the Renaissance, the Reformation, and the dawn of modern science.

Printing made possible the publication of annual astrological almanacs, which became wildly popular with ordinary people and flourished for the next three centuries. The almanacs contained forecasts for the year, advice on farming and health, indications of good and bad days for a variety of activities, and sometimes included ephemerides, which gave the positions of the planets in the signs of the zodiac for each day of the year, along with the phases of the moon.

Marsilio Ficino (1433-99), a Florentine priest, philosopher, and astrologer, was one of the most influential figures of the early Italian Renaissance. He translated pagan texts from Greek into Latin, including all the then-extant works of Plato as well as the works of the Neoplatonists and important Hermetic texts, and added his own comments. His mission was to reconcile pagan beliefs, including astrology and magic, with Christianity. As the vital link between medieval and Renaissance thought, Ficino was almost singlehandedly responsible for the revival of respect for pagan learning. Ficino's search for the one true religion that underlies all apparently contradictory beliefs was prophetic of New Age interfaith ideas, and his vision of the planets as psychological archetypes anticipated the ideas of Jung.

But the critics of astrology were already sharpening their weapons. In 1496 the Italian philosopher Pico della Mirandola (1463-94) published his *Disputationes adversus astrologiam divinatricem* (Treatise Against Predictive Astrology), a virulent attack that reduced astrology to prediction of the future. Like Augustine, Pico believed in a general planetary influence, but was violently opposed to anything that smacked of determinism. In Pico's eyes, astrology was

. . . the most infectious of all frauds . . . it corrupts philosophy, falsifies medicine, weakens religion, begets or strengthens superstition, encourages idolatry, destroys prudence, pollutes morality, defames heaven, and makes men unhappy, troubled, and uneasy.[100]

The fanatical Dominican monk Savonarola (1452-98) abhorred astrology as "an enemy of faith."

In response to its critics, astrologers launched a reform movement whose aim was to make astrology more astronomically accurate. According to Campion, it was the astrological reformers who were responsible for the astronomical discoveries of the next two hundred years. The greatest of these, of course, was that of Nicolas Copernicus (1473-1543), who proved that the sun was the center of the universe.

Although it has come to be associated with the Christian church, the Earth-centered cosmos was a Greek idea. Plato believed that the planets moved in perfect circles around the Earth, a theory that was contradicted by observation, since the planets obviously followed erratic orbits through the sky. Copernicus reasoned that if the Earth was not the center, it must be the sun, but his 1514 work, the *Commentariolus*, which contained this radical idea in its initial form, was not published until the nineteenth century. And his world-shattering *De revolutionibus orbium coelestium* (Concerning the Revolutions of the Heavenly Bodies) which contained the heliocentric theory in its final form, was not published until 1543, the year of his death. In light of the future policy of the Church, it is interesting to note that Copernicus received support for his theory from prominent Catholic bishops and cardinals, and even from Pope Clement VII.

Although Copernicus's world was sun-centered, the planets – including the Earth – still orbited in circles, à la Plato. Copernicus's universe, like Plato's, was beautiful and alive, and the purpose of his astronomy was to serve astrology, which in turn served the harmony of the state. Copernicus was deeply influenced by Hermetic ideas, especially the idea of the sun as the spiritual center of the universe. His cosmology, which is often regarded as a dramatic break with medieval superstition that heralded the triumph of science and reason, actually had deep roots in the past.

In 1517 the German priest Martin Luther (1483-1546) nailed his ninety-five theses opposing the corrupt practices of the Catholic church to the door of the castle church in Wittenberg. This event, which Luther's astrologer friend Philip Melanchthon rightly hailed as the "birthday" of the Protestant Reformation, launched a theological debate – and a series of religious wars – that divided Europe for centuries, and that coincided with the reform movement in cosmology. Luther believed that astrology came from God, and was therefore infallible, but that astrologers themselves were fallible:

I consider the rational basis of His celestial art as right, while the art itself is uncertain. That is, the signs in heaven and on Earth do not fail. They are the work of God and the angels.[101]

Luther opposed judicial astrology, as did Calvin. The term "judicial astrology" was used in the Middle Ages and early Renaissance to refer to the type of predictive astrology that was considered heretical by the Church. Judicial astrology was distinguished from "natural astrology" – e.g., medical and meteorological astrology, which were seen as acceptable because they were a part of the "natural philosophy" of the time. Natural philosophy was the term used for the study of nature and the physical universe before the advent of modern science.

Although he had very little to do with these cosmological debates, this history would not be complete without mention of Michel de Nostredame, better known by his Latin name, Nostradamus (1503-1566), probably the most famous astrologer of all time. His astrological almanacs were popular, but he is best known for his book of prophetic quatrains, *Les Prophéties*, which has been almost continuously in print since its first appearance in 1555. This, in spite of the fact that the four-line poems are almost completely incomprehensible.

Legend has it that while traveling in Italy Nostradamus knelt down before a young monk who went on to become Pope Sixtus V in 1585, twenty-one years after Nostradamus' death. After he became pope, one of Sixtus' first acts was to issue *Coeli et Terrae* (Heaven and Earth), the first Papal Bull forbidding the casting of horoscopes. A similar Bull was issued in 1631 by Pope Urban VIII, and papal opposition to astrology encouraged the censors to place astrological works on the Index of prohibited books. The popes continued to condemn astrology publicly but many of them used it privately, and some even cast horoscopes themselves.

In Elizabethan England, the plays of Shakespeare (1564-1616) are full of astrology. In *Lear*, Shakespeare puts a skeptical speech about astrology into the mouth of his villain, Edmund:

This is the excellent foppery of the world, that,
when we are sick in fortune – often the surfeit
of our own behavior – we make guilty of our
disasters the sun, the moon, and the stars: as
if we were villains by necessity; fools by
heavenly compulsion; knaves, thieves, and
treachers [traitors], by spherical predominance; drunkards,
liars, and adulterers, by an enforced obedience of
planetary influence; and all that we are evil in,
by a divine thrusting on: an admirable evasion
of whoremaster man, to lay his goatish
disposition to the charge of a star! My
father compounded with my mother under the
dragon's tail; and my nativity was under Ursa
major; so that it follows, I am rough and
lecherous. Tut, I should have been that I am,
had the maidenliest star in the firmament
twinkled on my bastardizing.

And in *Julius Caesar* he has the arch-conspirator Cassius remark:

Men at some time are masters of their fates:
The fault, dear Brutus, is not in our stars,
But in ourselves, that we are underlings.

We have no way of knowing what Shakespeare himself believed; but like Chaucer, he used astrology as a literary device in a way that proves it was part of the very fabric of life in his age.

5) The Seventeenth Century

And now we come to the great seventeenth century, the age that Arthur Koestler, in his flawed history of modern astronomy *The Sleepwalkers*, calls "the watershed": the turning point, the age when the sleepwalkers awoke and reason triumphed.

That reason triumphed is debatable; but what is not debatable is that the seventeenth century was an age of giants in art and science. It was the age of Donne and Milton and the Metaphysical poets in literature; of Rembrandt and Vermeer in art; and in astronomy – which by century's end was divorced from its dishonored partner, astrology – the age of Brahe and Kepler and Galileo.

Although he is less well known than his student, Kepler, Tycho Brahe (1546-1601) is regarded as the greatest observational astronomer since Hipparchus, the Greek astronomer of the second century BCE. Unlike Kepler, who had to struggle to make a living, Tycho had rich patrons, including the king of Denmark, and was able to build a magnificent observatory on the island of Hven near Copenhagen. Like Kepler, Tycho was a practicing astrologer of the reformer persuasion, and his goal as an astronomer was to serve astrology by arriving at more accurate calculations of the positions of the planets.

Tycho made an extraordinary discovery for the history of science when, in 1572, he saw a nova (literally, a new star) in the constellation Cassiopeia. According to Aristotelian cosmology, nothing new could occur in the divine region of the heavens beyond the moon, where all was supposedly perfect and unchanging. But the new star was obviously located beyond the moon. According to Campion,

It is difficult to overestimate the critical nature of this single fact and its implications were to be as great as Einstein's theory of relativity was for the twentieth century. An old certainty had been destroyed and the world would never be the same again.[102]

When the Danish king Frederick II died in 1588, Tycho was forced to find another patron. In 1599 he became "Imperial Mathematician" (i.e., both astrologer and astronomer) to the Holy Roman Emperor Rudolf II in Prague. At this point in his life – he died only two years later, in 1601 – he was most interested in continuing his work on the observation of the planets, particularly of the planet Mars. To assist him in this work he hired a young man who had recently published an impressive book on geometry and cosmology: the twenty-five-year-old Johannes Kepler.

Kepler (1571-1630) was a German Lutheran who was early exposed to the Copernican heliocentric view. His mystical streak is evident in the title of his first published work: *Mysterium Cosmographicum* (The Cosmic Mystery).

Tycho put his new assistant to work studying the motion of Mars. Kepler discovered that the orbits of the planets were not circular, as in the Platonic and Aristotelian universe. In 1605 Kepler developed the first two of his three laws of planetary motion governing the elliptical orbits of the planets around the Sun. In 1609 he published the *Astronomia Nova* (The New Astronomy), putting an end to the ancient idea of circular orbits which had persisted for two thousand years.

Kepler arrived at his third law of motion through a mystical vision of the Harmony of the Spheres. These ideas found their greatest expression in his 1619 work, *Harmonici Mundi* (World Harmony). Taken altogether, his three great laws of planetary motion anticipated and laid the foundation for Newton's law of universal gravitation.

Kepler had needed Tycho Brahe's observations to confirm what he got by intuition and arrive at his laws of planetary motion. With Tycho's help, Kepler discovered "the architecture of the solar system."[103] Immanuel Kant called Kepler "the most acute thinker ever born."

He was a man of contradictions. The founder of modern astronomy, he was also a devout Christian and an equally devout astrologer – a fact that has bedeviled historians of science ever since. Max Caspar, in the preface to his masterful biography of Kepler, apologizes for the amount of space he has granted to discussing his "astrological activities." To his credit, the dean of Kepler scholars states, "He who believes it possible to pass over them, with a few kind apologies, distorts the picture. Here it is a question not of our opinions on the subject, but his."[104]

Astrology was inseparable from Kepler's spiritual beliefs, which were in turn influenced by his early immersion in Pythagorean and Neoplatonic ideas. Kepler may have been the last great thinker for whom there was no divide between science and religion, or between astronomy and astrology:

Philosophy, and therefore genuine astrology, is a testimony of God's works, and is therefore holy. It is by no means a frivolous thing. And I, for my part, do not wish to dishonor it.[105]

Note that "*genuine* astrology:" Like his teacher Tycho Brahe, Kepler was critical of the kind of astrology practiced by his contemporaries and very much in favor of reform:

Though many of the rules in this Arabic art amount to nothing, still all that which therein is contained is not nothing, and may therefore not be discarded . . .; rather one must separate the gems from the slag.[106]

Kepler believed that there was a core of validity in the ancient science, and in one of his scientific treatises, he warns

certain Theologians, Physicians, and Philosophers . . . that while justly rejecting the stargazers' superstitions, they should not throw the baby out with the bathwater.[107]

Kepler believed that the highest goal of astrology (and indeed of all science) was knowledge of God, and he hated to see the divine science abused by astrologers for trivial or vulgar purposes:

finding lost objects or determining whether a wife was unfaithful. He also objected to precise prediction. His forecasts tended to speak in terms of probabilities rather than certainties.

In "separating the gems from the slag" Kepler threw out the majority of the rules that had come down from the ancients, including the signs of the zodiac; the doctrine of rulerships (i.e., that the sign Aries is ruled by the planet Mars, and so on); the meanings of the twelve houses into which a chart was divided; the traditional personalities of the planets; and several other items in the standard toolbox of the medieval astrologer.

His pared-down version of astrology was limited to the planets and their aspects – the angular relationships between the planets, which he saw as related to the musical intervals. He also retained the significance of the basic angles of the horoscope, that is, the Ascendant and the Mid-Heaven, and of the aspects of the planets to those two points. He retained the aspects of the planets because they had a sound, measurable foundation in astronomy, whereas the signs and the houses were mental constructs.[108]

Kepler actually discovered several aspects beyond the Ptolemaic five (conjunction, opposition, square, trine, and sextile) that are related to harmonics and are still used by modern astrologers. The Keplerian aspects include the quintile and biquintile, aspects of 72 and 144 degrees, which divide the circle by five; the decile, an aspect of 36 degrees, which divides the circle by ten; and the sesquiquadrate, an aspect of 135 degrees, in other words, a square (90 degrees) and a half.[109]

Ironically, he also invented secondary progressions, a system of prediction which many modern astrologers find indispensable in examining future trends for their clients.

While Kepler disliked precise predicting and condemned the superstitious debasement of astrology by his contemporaries, he did believe in an *effectus generalis*: a general influence of the planets on nature, and on the character of human beings at the time of their birth:

> . . . the person in the first igniting of his life when he now lives for himself and can no longer remain in the maternal body receives a character and pattern of all the constellations of the heaven or of the form of the rays flowing onto the Earth which he retains until he enters his grave.[110]

Since he was perennially short of funds, the standard view of historians of science has been that Kepler cast horoscopes for money without believing in astrology or respecting it himself, but this is a misreading of his character. He famously wrote that:

The belief in the effect of the constellations derives in the first place from experience, which is so convincing that it can be denied only by those who have not examined it.[111]

And toward the end of his life he wrote:

A most unfailing experience (as far as can be expected in nature) of the excitement of sublunary natures by the conjunctions and aspects of the planets has instructed and compelled my unwilling belief.[112]

    Kepler cast horoscopes for himself and for the members of his family all of his life. He even drew up a conception chart for himself.[113] On his wedding day, he noted in his diary that he married his first wife *calamitoso coelo* (beneath an ominous sky), a forecast that turned out to be all too accurate. (It was not a particularly happy marriage.) At the beginning of each year of his life he examined the aspects formed by transits of the planets to their positions in his birth chart. For his sixtieth year he noted that some of the planets would repeat their natal positions, a concurrence that is sometimes found in death charts. He did indeed die at noon on November 15, 1630, at the age of 60.

    Although Kepler was the greatest of the astrological reformers, his efforts to simplify and improve astrology ultimately failed: it was the fatalistic version of popular astrology featured in the almanacs that prevailed. But Kepler's profound reverence for serious astrology anticipated that of the spiritually oriented humanist astrologers who led the renaissance of the mid-twentieth century. Like Kepler, they believed that the aspects of the planets provided clues to an individual's character and personality, and even to their soul. In this sense astrology was and is spiritual work.

    Kepler published his first two laws of planetary motion in 1609, the same year that Galileo looked in his telescope and saw the four moons of Jupiter. Campion considers the years

1609-10 the beginning of the modern age.

Galileo (1564-1642) was born the same year as Shakespeare. Like his contemporary, Kepler, he started out as a teacher of mathematics, a *mathematicus*, which in those days meant someone versed not only in mathematics but in the related fields of astronomy and astrology, which were still a single science. He taught first at the University of Pisa, the town where he was born, and later at the University of Padua.

In 1610 he published his *Siderius Nuncius* (The Starry Messenger) in which he named the four moons that he had observed orbiting Jupiter the "Medicean stars," in honor of the Medicis, the wealthy banking family of Florence. A clear bid for patronage, it worked: later that year, Cosimo de Medici appointed Galileo court mathematician and astrologer.

Galileo discovered that the Earth's moon, which had heretofore been regarded as a perfect sphere, had mountains and craters, just like the Earth, and that Venus had phases like the moon, which proved that Venus orbited the sun, not the Earth. All of these discoveries lent support to the Copernican heliocentric model of the universe, which Galileo had initially doubted. Now he became a convert to the sun-centered world of Copernicus and in 1613 he began publicly defending it in his *Letters on Sunspots*.

These discoveries rocked the whole educated world of Europe. In England in 1611 the poet John Donne wrote in his poem "An Anatomy of the World:"

And new philosophy calls all in doubt
The element of fire is quite put out
The sun is lost, and th'Earth, and no man's wit
Can well direct him where to look for it.
And freely men confess that this world's spent
When in the planets, and the firmament
They seek so many new. . . .
'Tis all in pieces, all coherence gone. . . .

Galileo saw no contradiction between the Sun-centered view and the Bible, which he did not take literally; but the Church became alarmed, not so much by the heliocentric view itself as by the threat to their authority: only the Church was allowed to pass judgment on these matters.

They suspended Copernicus's 1543 *De Revolutionibus orbium coelestium* (On the Revolutions of the Heavenly Spheres) pending removal of those passages that implied that the Earth moved like a planet.

It is well known that Galileo was eventually tried by the Inquisition, placed under house arrest, and instructed not to teach the heliocentric view. What is not so well known is that Galileo had an earlier run-in with the Church for practicing – of all things – astrology.

Although until recently there has been next to nothing about it in mainstream histories of science, Galileo was a practicing astrologer for a significant portion of his long career. While he was living in Padua and teaching at the University there he cast horoscopes for money. His account book lists a number of clients who paid him sixty lire *per sortem* (for each "lot" or "destiny").

It was during this period that Galileo's mother asked his private secretary, a man named Silvestro Pagnoni who lived in Galileo's house, to spy on him. Pagnoni told the Inquisition in Venice that in the eighteen months he lived under Galileo's roof he never once saw him go to Mass, but that he did see him draw up numerous "nativities" (birth charts) upon which he made "judgments" (predictions). The charge brought against Galileo was not that he practiced astrology, which was a normal activity for a *mathematicus* in those days, but that he practiced a deterministic or fatalistic astrology that seemed to rule out free will. In the eyes of the Church, this was heresy.

In the end the charges were dropped, but the incident sheds light on Galileo's practice of astrology, which was quite different from Kepler's. Galileo made use of the whole panoply of traditional astrology: signs, houses, rulerships, etc., and did not hesitate to predict outcomes. Galileo was not a reformer – at least, not of astrology.

Galileo's library, which was reconstructed by Italian historian Antonio Favaro in the late nineteenth century, contained numerous astrological texts with marginal notes in his handwriting. These books include an ephemeris for 1598-1610, with a long astrological introduction; a Latin translation of Ptolemy's *Tetrabiblos*; and an astrological work by Tommaso Campanella, a contemporary of Galileo's, in which Galileo marked all references to himself.[114]

That Galileo did not cast horoscopes for purely financial reasons, or to curry favor with rich patrons like Cosimo de Medici, there is ample evidence. In the Galileo archives in the Biblioteca Nazionale Centrale in Florence there still exists what historian Darrel Rutkin, who wrote his 2002 doctoral thesis at Indiana University on the history of astrology, calls "a grossly understudied document of striking interest."[115]

Ms. Gal. 81 contains fifty pages of horoscopes in Galileo's handwriting. Among these are horoscopes for himself and his daughters and brief judgments on seven, including those of Virginia and Livia, whose character and personality traits he attributes to the influence of the planets. For example, of his older daughter Virginia (who became Sister Maria Celeste) he writes, "Mercury endowed with many dignities promises a fine mind." Rutkin describes this manuscript as "the working notebook of a practicing astrologer."[116]

Finally, Galileo's letters are full of astrology. While some are to patrons, others are to friends, clients, and fellow astrologers. Early in 1609, the Grand Duchess Christina, wife of the Grand Duke Ferdinand and mother of Prince Cosimo, asked Galileo to determine an accurate date of birth for her husband, who was gravely ill, to aid in a medical prognosis. To do this required a knowledge of the complicated process of rectification. That Galileo took on this assignment shows that he had a detailed knowledge of astrology. In a letter dated January 16, 1609, he gives the requested determination.

Galileo carried on an extensive correspondence with Ottavio Brenzoni, a Veronese physician who consulted him for astrological assistance in medical diagnosis and treatment.[117]

Of particular interest is a long letter Galileo wrote in 1611 to a friend who took issue with certain statements he had made in the *Sidereus nuncius*, because in this letter Galileo gives his views on how astrology works. He believes that not only light, but motion, is involved, and his defense of the latter mechanism betrays a level of understanding of astrology that only comes with deep study and considerable experience.[118]

According to Rutkin, as late as 1618 Galileo was receiving requests for nativities and judgments. Later letters in the archive have been cited to prove that Galileo eventually rejected astrology; but he was simply objecting to certain practices on the part of his contemporaries.

Galileo had been an evangelist for the heliocentric system of Copernicus since his *Letters on Sunspots* of 1613. In his 1632 *Dialogue on the Two Chief World Systems,* three characters debate the heliocentric vs. the geocentric system, but it is clear which of them the author believes is right. On April 11, 1633 Galileo was put on trial by the Inquisition for supporting the theory that the Earth revolves around the Sun. Later that month he agreed to plead guilty and was placed under house arrest in his home in Florence for an unlimited period that turned out to be the rest of his life. He was instructed not to teach the heliocentric view. He continued to study and to write, and his *Discourse on the Two New Sciences* (1638) laid the foundations of modern physics.

He gradually went blind and died in 1642, the year that Newton was born. It wasn't until 1992 that the Catholic church formally admitted that Galileo's views on the solar system were correct.

Galileo's astrology has been ignored by most historians of science. Kepler's astrology and Newton's alchemy have been tolerated, but it has been felt intolerable that the father of modern physics, the heroic champion of the scientific revolution who was persecuted by the Church for his beliefs, was an astrologer.[119]

For example, the official Italian edition of Galileo's works reduces the fifty manuscript pages containing his hand-drawn horoscopes in the Biblioteca Nazionale Centrale in Florence to two.[120] According to Rutkin, this astrological manuscript of Galileo's has been known at least since the publication of Antonio Favaro's *Galileo Galilei e lo Studio di Padova* in 1883, but Favaro's work on Galileo had to wait until 1993 for an English translation.

And yet an otherwise excellent scholar like Charles B. Schmitt, author of *Studies in Renaissance Philosophy and Science (*1972*),* can conclude that "at least Galileo was not attracted to occult mathematics," i.e., astrology. Rutkin comments, "Such a statement by a profound and open-minded scholar clearly indicates an important blind spot in the historiography of science in general, and of Galileo in particular."[121]

In the 1998 *Cambridge Companion to Galileo*, astrology is completely ignored.[122]

As a consequence of this blind spot, Galileo is one of the least understood figures in the history of science. Victim of religious superstition in his time, he is the victim of materialist projection in ours. Galileo may not have been as brilliant an astrologer as Kepler, but an astrologer he was, and to ignore this fact is to misunderstand both Galileo and astrology, and to falsify science.

The seventeenth-century English philosopher and scientist Francis Bacon (1561-1626), regarded as the father of the scientific method, called for an *astrologia sana*, an astrology purged of superstition. Author of the *Novum Organum* (A New Method, 1620), he believed in the influence of the planets, but had no use for astrology as it was practiced in his day. Bacon was a reformer in the tradition of Kepler:

As for Astrology . . . I would rather have it purified than altogether rejected. . . . The last rule (which has always been held by the wiser astrologers) is that there is no fatal necessity in the stars; but that they rather incline than compel. We will add one thing more (wherein I shall certainly seem to take part with astrology, if it were reformed); that we are certain the celestial bodies have other influences besides light and heat.[123]
v

In his 1623 work *De augmentis scientiarum* (Concerning the Improvement of the Sciences), Bacon advocated an empirical approach to "revolutions" (i.e., mundane astrology – the effects of the planets on life on Earth in general, rather than on individuals):

Astrologians (if they be not wanting to their Profession) may make a collection from the faithfull reports of History, of all greater contingences; as Inundations, Pestilences, Warres; Seditions; and . . . the deaths of Kings: and may contemplate the situation of the Heavens . . . according to those general rules which we have already set downe; to know in what postures the Heaven were, at those times, when such effects came to passe; that so[where] there is a cleere, and evident consent, and concurrence of events; there a probable rule of Prediction may be inferred.[124]

The important English astrologer William Lilly (1602-1681) continued the tradition of attempting to reconcile astrology with Christianity. His *Christian Astrology* (1647) was the first astrological text to be published in English. His annual almanacs were popular, and were translated into Dutch, German, Swedish, and Danish. He specialized in horary astrology, and an amazing percentage of his predictions came true. In 1648 he predicted not only the plague that would decimate London in 1665 but the Great Fire of 1666 that nearly destroyed it. Lilly was not a reformer, and it was his astrology – not Kepler's pared down version – that survived into the twenty-first century.

The French astrologer Jean-Baptiste Morin (1583-1656) was also skilled at prediction. Court astrologer to the kings and queens of France and to Cardinal Richelieu (whose death he predicted within ten hours), in 1638 he waited concealed in the queen's bedroom to record the exact moment of the birth of the future King Louis XIV. He predicted the exact dates of the death of Louis XIII, Wallenstein, and Gustavus Adolphus.[125]

In spite of its popularity with ordinary people and its occasional extraordinary successes among the great, astrology was increasingly an object of contempt for the writers of the day. Samuel Pepys ridiculed Lilly in his famous diary, and the great English writer Jonathan Swift (1667-1745) satirized astrology in *A Tale of a Tub: The Battle of the Books and Other Satires* (1704).

The English poet John Dryden (1631-1700) is an interesting case. In 1668 he ridiculed astrology in his play *An Evening's Love: or, the Mock Astrologer*; but his poem "Song for St. Cecilia's Day" (1687) eloquently expresses the medieval world view on which astrology is based. Here are the first and last stanzas:

From Harmony, from heav'nly Harmony
This universal frame began:
When Nature underneath a heap
Of jarring Atomes lay,
And could not heave her Head,
The tuneful Voice was heard from high:
Arise, ye more than dead.
Then cold and hot and moist and dry
In order to their Stations leap,
And Musick's pow'r obey.

As from the Pow'r of Sacred Lays

The spheres began to move,
And sung the great Creator's Praise
To all the bless'd above;
So, when the last and dreadful Hour
This crumbling Pageant shall devour,
The Trumpet shall be heard on high,
The dead shall live, the living die,
And Musick shall untune the sky.

In fact, Dryden was himself an astrologer. He cast a horoscope for the birth of his son Charles and accurately predicted the events of his short and difficult life.[126]

The ambivalent attitude toward astrology that characterized the late seventeenth century is nowhere better illustrated than in the figure of John Flamsteed (1646-1719), first Astronomer Royal at the Greenwich Observatory in England. Darrel Rutkin writes that after completing an intensive critique of astrology in 1673 (which he never published), he cast a horoscope for the foundation of the Greenwich Observatory in 1675:

Indeed, he and Edmund Halley [best known for computing the orbit of the eponymous comet] both provided the most accurate astronomical data available for the tables printed in George Parker's popular astrological almanacs in the 1690s.[127]

The almanacs continued to be published until the end of the eighteenth century. Benjamin Franklin's *Poor Richard's Almanack* (published continuously from 1732 to 1758) contained, according to the cover of the issue for 1739):

Lunations, Eclipses, Judgment of the Weather, Spring Tides, Planets' Motions & Mutual Aspects, Sun and Moon's Rising and Setting, Length of Days, Time of High Water, Fairs, Courts, and observable Days. . .[128]

along with poems, proverbs, puzzles, and practical household hints.

Serious astrology continued to have its practitioners: Nicholas Culpeper (1616-1654), a distinguished English herbalist, physician, and astrologer, famously remarked, "A medical man without astrology is like a lamp without oil."[129] And in 1686 John Goad (1615-1689), an English meteorologist, published *Astro-Meteorologica*, an attempt to establish a scientific system of weather prediction based on the aspects of the planets. In this he anticipated the work of twentieth-century American meteorologist John Nelson (1903-1984).[130]

But increasingly common as the seventeenth century wore on was the attitude of influential French textbook author, Jacques Rohault, who devoted a chapter to astrology in his 1671 *Traité de Physique* (Treatise on Physics). Rohault rejected astrology on logical grounds and concluded, "Not to insist any longer upon this Subject, which does not deserve to have any more said of it, and which is not worth being seriously treated by any Philosopher."[131]

Substitute the word "scientist" for "philosopher," and this sentence might have appeared in a recent issue of the *Humanist*. It is the voice of modern science.

And the voice of modern science is the legacy, above all, of René Descartes (1596-1650), the father of modern rational materialism. In contrast to the centuries-old cosmology of Plato and Aristotle, he saw no order or purpose in the cosmos. In his *Discourse on Method* (1637), he called astrology a "false science" and put it in the same category as alchemy and magic. Mainstream scientists have followed his example ever since.

By the end of the seventeenth century astrology was, as Campion puts it, "excluded from educated discourse." The revolution in astronomy had motivated the reformers of astrology, who had begun speaking in terms of probabilities rather than certainties. They avoided precise prediction of future events, but their efforts at reform were largely ignored. The educated classes, excited by the new discoveries of science, lost interest in astrology. The almanacs were still popular with the masses, but with the notable exception of Goethe, no leading Western intellectual took astrology seriously until Jung.

According to Campion, the underlying causes of this major paradigm shift are still unknown. There is no logical reason why the views of Copernicus should have destroyed astrology. Contrary to the statements of the editors of the *Humanist* (and to the opinions of most modern astronomers), astrology has never been disproved.

The great Christian hymn, "The Spacious Firmament on High," by the eighteenth-century

English poet and essayist Joseph Addison, which was set to music by Franz Josef Haydn, declares:

The spacious firmament on high
With all the blue ethereal sky
And spangled heav'ns a shining frame

Their great Original proclaim.
Th'unwearied sun, from day to day,
Does his Creator's power display,
And publishes to every land,
The work of an almighty hand.

Soon as the evening shades prevail,
The moon takes up the wondrous tale,
And nightly to the list'ning earth
Repeats the story of her birth;
Whilst all the stars that round her burn,
And all the planets in their turn,
Confirm the tidings as they roll,
And spread the truth from pole to pole.

What though, in solemn silence, all
Move round this dark terrestrial ball;
What though no real voice, nor sound,
Amidst their radiant orbs be found;
In reason's ear they all rejoice,
And utter forth a glorious voice
Forever singing as they shine;
"The hand that made us is divine."

This beautiful hymn is a splendid example of the old idea – at least as old as the Greeks of the fifth century BCE – of the "argument from design:" the idea that the elegance and complexity of the universe proves the existence of God.

In the eighteenth century, God is still part of the equation and the universe still has purpose and meaning. But let's look more closely: Although for poetic purposes the eighteenth-century author retains the ancient geocentric view that the stars and planets "move round this dark terrestrial ball," he no longer accepts a literal version of the pagan superstition of the Music of the Spheres. All is now silence. The glorious voice is only heard "in reason's ear." The disenchantment of the universe has begun.

6) The Eighteenth Century

Isaac Newton (1642-1727) studied the laws of motion, the shape of the Earth, and the tides, and discovered the law of universal gravitation. In his momentous three-volume work *Philosophia Naturalis Principia Mathematica* (1687: Mathematical Principles of Natural Philosophy), normally known simply as the *Principia*, he wrote that every particle of matter in the universe attracts every other particle with a force varying inversely as the square of the distance between them and directly as the product of the masses of the two particles.

Newton's study of gravity led him to an understanding of the flattening of the Earth's poles and the tilting of the Earth on its axis, and thus to the cause of the precession of the equinoxes originally discovered by Hipparchus in 150 BCE.; but Newton's astronomy did not lead him to an acceptance of astrology. The oft-cited legend that Newton told Edmond Halley, who expressed doubts about astrology, "Sir, I have studied it, you have not," is probably apocryphal.

Newton was an enthusiastic alchemist, but like Descartes, he rejected astrology along with the Neoplatonic idea of a living universe. But unlike Descartes, Newton did believe in God, and in a universe filled with meaning and purpose:

This most beautiful system of the Sun, planets, and comets could only proceed from the counsel and dominion of an intelligent and powerful Being.[132]

For example, Newton believed that comets were messages from God. But Newton's universe came to be seen as mechanistic, lifeless, random, and without meaning.

In spite of the efforts toward reform in the sixteenth and seventeenth centuries, astrology lost its legitimacy in the eighteenth century in a process that is not yet fully understood.[133]

Astrology was outlawed in England in 1736 under the Witchcraft Act, and again in 1824 under the Vagrancy Act.

Eustachio Manfredi, author of an Italian ephemeris for 1715-1725, stated that he was deliberately removing "the stain, astrology, from his ephemeris, unlike his predecessors Regiomontanus, Magini, and Kepler. Only tables of planetary aspects remained."[134]

And Ephraim Chambers, in the preface to his popular 1728 *Cyclopedia*, made a distinction between natural astrology, by which he meant the acceptable influence of the planets on the weather and in a general sense, on human beings, and judicial astrology, which he dismissed as superstition. But rather than rejecting astrology altogether, he called for its reform.

Chambers' discussion of astrology was translated into French and appeared in the first (1751) edition of the famed *Encyclopédie* of Denis Diderot and Jean d'Alembert. Rutkin contrasts Chambers' reforming attitude to that of the authors of the first (1768) edition of the *Encyclopaedia Britannica*, whose definition of astrology reads as follows:

A conjectural science, which teaches to judge of the effects and influences of the stars, and to foretel [*sic*] future events by the situation and different aspects of the heavenly bodies. This science has long ago become a just subject of contempt and ridicule.[135]

This rejection of astrology was academic but not popular. Astrological texts – especially the ever-popular almanacs – continued to be translated into English.

Even the legendary lover Giacomo Casanova (1725-98) took aim at astrology in his memoirs:

A curious fancy increased my delight, namely, the thought of becoming a famous astrologer in an age when reason and science had so justly demolished astrology. I enjoyed the thought of seeing myself sought out by crowned heads, which are always the more accessible to superstitious notions.[136]

On the other hand, Johann Wolfgang von Goethe (1749-1832), generally regarded as the greatest mind of his age, begins his autobiography as follows:

On the 28th of August, 1749, at mid-day, as the clock struck twelve, I came into the world, at Frankfort-on-the-Main. My horoscope was propitious: the sun stood in the sign of the Virgin, and had culminated for the day; Jupiter and Venus looked on him with a friendly eye, and Mercury not adversely; while Saturn and Mars kept themselves indifferent; the moon alone, just full, exerted the power of her reflection all the more, as she had then reached her planetary hour. She opposed herself, therefore, to my birth, which could not be accomplished until this hour was passed.[137]

Clearly, Goethe was an exception to the general trend among the educated classes.

By the end of the eighteenth century astrology was opposed by the scientific establishment as well as by its perennial enemy, the Christian church: a double dose of disdain that continues to this day (and a rare example of a subject on which science and religion are in agreement).

Astrologers who continued to cast horoscopes operated in a vacuum, without social or professional support. Intellectuals believed in neither signs nor influences, let alone precise predictions. Nothing was real to them but what was once called "natural astrology:" the effects of the moon on the tides and the sun on the seasons. The separation of astrology from astronomy was complete.

And yet, the discovery of the planet Uranus by the German-born British astronomer Sir William Herschel on March 13, 1781 was an astrological, as well as an astronomical, event. Herschel originally wanted to name the planet in honor of his patron, King George III of England, which would not have gone over very well in the American colonies, where the people had some familiarity with astronomy and astrology through the popular almanacs.

The first of the so-called modern planets to be discovered, Uranus could only have been seen after the invention of the telescope, since it is barely visible to the naked eye. Astrologers had to start collecting data to determine its eventual use. Here is *Larousse* on Uranus:

The attributes of the modern planets have been deduced by careful observation, over a period of time, of cultural trends that appeared around the time of their discovery. Uranus first impinged on human consciousness around the time of the American Revolution, the French Revolution, the harnessing of electricity, the Industrial Revolution, and numerous inventions of far-reaching consequences, such as steam-driven machinery. . . .. Astrologers have come to associate this planet with revolution, invention, innovation, sudden change, the unexpected.[138]

When Uranus was first discovered, perhaps because of the violence of some of the changes that coincided with its appearance, many astrologers regarded the planet as "malefic" – doing harm.

There's a funny story about an English astrologer named John Varley who was in the habit of checking the transits of the planets to his natal chart in an ephemeris at the beginning of every day. If the aspects were bad he would refuse to leave the house.

On June 21, 1825 Varley saw that the position of Uranus a few minutes before noon indicated the possibility of sudden danger. He had his son take his pocket watch to the watchmaker in Regent Street and have it set precisely to Greenwich time. His son returned to find his father pacing up and down, increasingly agitated as noon approached. Could he have made a mistake?

Varley sits down to check his calculations, when suddenly there is a cry of Fire! from outside the building. He and his son rush out only to discover that their house is in flames. But Varley goes back inside and sits down at his desk to write an account of his discovery. In the end he loses his home and all of his possessions, but he is delighted, because it proves that he was right about Uranus! To fully appreciate this story, I think you have to be an astrologer. . ..[139]

7) The Nineteenth Century

In the nineteenth century there was a gradual revival of horoscopic astrology, especially in England. This revival featured some eccentric characters who practiced and published under pseudonyms, many the names of Hebrew angels: Raphael, Zadkiel, Sepharial, and Aphorel, among others.

These grandiose pseudonyms may have been motivated partly by fear: Early in the century astrology had become criminalized under an old Georgian act "for the punishment of Idle and Disorderly Persons, Rogues and Vagabonds." Section 4 of the Vagrancy Act of 1824 applied to every Person pretending or professing to tell Fortunes, or using any subtle Craft, Means, or Device, by Palmistry or otherwise, to deceive and impose on any of His Majesty's Subjects.[140]

The police used paid informers to entrap unsuspecting astrologers by posing as clients. Prison sentences could be up to three months with hard labor, usually with no possibility of appeal.

In 1813 an astrologer named Thomas White was arrested and his books and papers seized. He was convicted under the Vagrancy Act and died after three months in Winchester gaol. In 1844 J. Bradshaw was convicted of fortune telling and imprisoned for a month, and between 1851 and 1853 there were four more prosecutions of astrologers.[141]

The English poet William Blake (1757-1827) was indignant at these arrests. He became the center of a group of artists who were fascinated by astrology, clairvoyance, spiritualism, and other occult teachings.

An astrologer who called himself Zadkiel (Richard James Morrison, 1795-1874) fought for years to get Section 4 of the Vagrancy Act repealed. Among his foes was the violently anti-astrological Society for the Diffusion of Useful Knowledge (SDUK), founded in 1825, a Victorian precursor to the debunking twentieth-century Committee for the Scientific Investigation of Claims of the Paranormal (CSICOP).[142]

Zadkiel's efforts to make astrology respectable (or at least legal) failed, but his annual almanacs sold well throughout the nineteenth century and into the twentieth. But his almanac for 1861 precipitated a major event in his life. In it he included a dire prediction regarding the health of Prince Albert, the beloved and popular consort of Queen Victoria, who fulfilled it by dying suddenly on December 14. This accurate prediction brought an avalanche of abuse down on Zadkiel's (Morrison's) head. In 1862 the *Daily Telegraph* commented:

There is a fellow who calls himself Zadkiel, and who for thirty-two years, it seems, has been suffered to publish annually a farrago of wretched trash which he calls an almanac, and in which, pretending to interpret the "voice of the stars," he gives vent to a mass of predictions on public affairs.[143]

The *Times* excoriated astrology as

a passion, like table-turning or spirit-rapping in our own time, chiefly of those to whom morbid excitement had become a necessity: silly women, worn-out fashionables, and unprincipled adventurers.[144]

A letter appeared in the *Daily Telegraph* accusing Zadkiel of perpetrating a deliberate fraud during sessions with a crystal ball. Zadkiel hired a lawyer and sued for libel. After the long trial that ensued Zadkiel actually won his case, but it was a Pyrrhic victory: astrology was not vindicated. The Vagrancy Act remained on the books until 1989, when it was quietly repealed along with some other statutes that had long since ceased to be enforced.

Zadkiel's colleague Raphael (Robert Cross Smith, 1795-1832) in his *Manual of Astrology* (1828), gives a famous example of astrological twins: persons born on the same day in roughly the same place, and who hence have almost identical horoscopes. King George III and an ironmonger named Samuel Hemmings were both born June 4, 1738. They were said to look alike; Hemmings went into business the same month George succeeded to the throne; both men were married on September 8, 1761, and both died on January 29, 1820.[145]

The title page of Raphael's *Manual of Astrology* reads:

A Manual of Astrology, or The Book of the Stars, being the art of foretelling Future Events, *by the influences of the Heavenly Bodies*, In a manner unattempted by any former Author and divested of the Superstitions of the Dark Ages. By Raphael, *The Author of 'The Astrologer of the Nineteenth Century,' 'The Prophetic Messenger,' &c. &c.* "The Book of past Times shall be unsealed." –Ancient Prophecy[146]

But Raphael is best known for his popular annual ephemeris, which continued to be published under the Raphael name by a succession of astrologers throughout the twentieth century. Zadkiel's almanac survived until 1931, but Raphael's was still going strong when I began studying astrology in 1969. Long before the age of computers I relied on *Raphael's Ephemeris*, reduced to a slim pamphlet but still crammed full of technical information, including everything needed to cast a horoscope. I still have my very first copy. The title page of the 1969 edition reads in part:

Raphael's Astronomical Ephemeris of the Planets' Places for 1969 with Tables of Houses for London, Liverpool & New York / Containing the Longitudes of all the Planets daily and their Latitudes and Declinations for every other day, with the Lunar and Mutual Aspects for every day, &c., &c. / A Complete Aspectarian / Ephemeris Time Observed Throughout /*Caution:* Refuse inaccurate imitation issues.

The planet Neptune was discovered in 1846 by mathematical prediction rather than observation, because of irregularities in the orbit of Uranus. As with the other modern planets, it has come to be associated with cultural trends around the time of its discovery. *Larousse* on Neptune:

The late nineteenth and early twentieth centuries saw the development of anaesthesia and chemotherapy; the birth of hypnotism and psychoanalysis; Freud's exploration of the unconscious and Jung's search for the collective unconscious; renewed interest in the occult and Oriental philosophy and the rise of spiritualism; the abolition of slavery, the end of serfdom in central Europe, and the rise of communism, internationalism, and pacifism; the publication of Einstein's theory of relativity and Max Planck's quantum theory, breaking down Newtonian concepts of space, time, and matter; the stream of consciousness movement in literature, cubism in art, and the birth of the cinema.[147]

The theories of Darwin and the rise of scientific materialism created a spiritual crisis and a need to explore the possibility of psychic phenomena, and especially of life after death. The English Society for Psychical Research was founded in 1882 and its American counterpart in 1885.

Neptune came to be associated with spiritualism, clairvoyance, psychic phenomena, mesmerism (later renamed hypnosis), utopianism, anaesthesia, and drugs. It came to rule dreams, illusion, delusion, fantasy, and film. The core meaning of Neptune is the dissolving of boundaries. Universal languages like mathematics and music are regarded as Neptunian. *Larousse* concludes that Neptune rules "all states in which ordinary categories of perception break down, conditioning is transcended, and there is receptivity – for good or ill – to other levels of reality."[148]

In 1875 the prolific Russian author and spiritual teacher Helena Petrovna Blavatsky

(1831-91) founded the Theosophical Society in New York. Theosophy means "divine wisdom." Its core belief was that all religions are one, and Madame Blavatsky was influential in introducing Eastern religions to the West. She was not an astrologer herself, but she respected astrology and created a safe environment for its practice. In her first major work, *Isis Unveiled*, she writes:

Astrology is to exact astronomy what psychology is to exact physiology. In astrology and psychology one has to step beyond the visible world of matter, and enter the Platonic and Aristotelian schools, and it is not in our century of Sadducean scepticism that the former will prevail over the latter.[149]

8) The Twentieth Century

At the beginning of the twentieth century astrology was outlawed in New York State. Article 165.35 of the New York Penal Code classifies fortune telling as a class B misdemeanor, and defines it as follows:

A person is guilty of fortune telling when, for a fee or compensation which he directly or indirectly solicits or receives, he claims or pretends to tell fortunes, or holds himself out as being able, by claimed or pretended use of occult powers, to answer questions or give advice on personal matters or to exorcise, influence or affect evil spirits or curses; except that this section does not apply to a person who engages in the afore-described conduct as part of a show or exhibition solely for the purpose of entertainment or amusement.[150]

In 1915 in New York City astrologer Evangeline Adams (1868-1932) was arrested for fortune telling. At her trial she expounded the principles of astrology and illustrated its practice by reading a blind chart that turned out to be that of the judge's son. The judge was so impressed by her character and intelligence that he ruled in her favor, concluding that "the defendant raises astrology to the dignity of an exact science."[151]

The Irish poet William Butler Yeats (1865-1939), one of the literary giants of the twentieth century and winner in 1923 of the Nobel Prize in literature, was an enthusiastic lifelong astrologer. In 1890 he joined the Hermetic Order of the Golden Dawn, a magical and esoteric society founded by English Freemasons, and later he became a member of the Theosophical Society. His wife, George Hyde-Lees, was a medium, and early versions of his book *A Vision* came out of their experiments in automatic writing. A related work entitled *Astrology and the Nature of Reality* was published after his death. The poet W. H. Auden revered Yeats but criticized this aspect of his work as the "deplorable spectacle of a grown man occupied with the mumbo-jumbo of magic and the nonsense of India."[152]

The Astrological Lodge of the Theosophical Society was founded in 1914. Madame Blavatsky's student Alice Bailey (1880-1949) wrote prolifically on occult subjects, including esoteric astrology, which she called "the astrology of the soul." Esoteric astrology was primarily concerned with spiritual evolution. From Theosophy it inherited a belief in Eastern ideas like reincarnation and karma, but it was also influenced by Christian theology. For example, the angles of the birth chart were sometimes referred to as "the cross of matter on which the native is crucified." Esoteric astrology – and astrology in general – eventually became a vital part of the constellation of ideas now referred to as New Age. Campion believes that Bailey is largely responsible for popularizing the term New Age as a synonym for the Age of Aquarius.

The astrological ages – periods of approximately 2,150 years when the sun rises against the same constellation at the vernal equinox – are the result of the phenomenon of precession of the equinoxes.[153] For example, the period from the time of Christ to the late twentieth century, when the vernal point occurred in the constellation of Pisces, is known as the Age of Pisces, or the Piscean Age. Since the vernal point moves backwards through the signs, the next astrological age will be the Age of Aquarius.

Among astrologers, opinions differed on exactly when the New Age was to begin, but all agreed that astrology had the potential to provide people with the self-understanding that would prepare them for the transformation to come. As an essential tool for personal growth, astrology came to be the *lingua franca* of the New Age.[154]

Rudolf Steiner (1861-1925) was a theosophist but left the Theosophical Society in 1912 and founded his own movement, Anthroposophy. He practiced biodynamic farming by the phases of the moon. His astrology was sidereal: i.e., he used the constellations instead of the signs of the zodiac. Like many modern astrologers, Steiner believed that the goal of astrology was spiritual: to find the purpose of a given incarnation. He had no interest in practical prediction.

The most influential of the astrologers of the Theosophical school was the Anglo-Scottish astrologer and author Alan Leo (1860-1917). Through his teaching and his prolific writing – he produced, either in whole or in part, some thirty books, and founded the magazine *Modern Astrology* – he almost singlehandedly created modern astrology. It was largely through his influence that oriental ideas like reincarnation, karma, and the journey of the soul became part of modern astrology. A majority of modern astrologers now believe in reincarnation and karma, and many of them avoid precise prediction and focus on self-knowledge and the spiritual path.

Leo's delineations of birth charts provided clues to past lives. The mantra of his esoteric astrology was the idea – as old as Heraclitus – that "character determines destiny." He was opposed to fatalistic predictive astrology; but ironically, he was twice prosecuted for fortune telling. A born leader and organizer, Leo invented a way to market and mass-produce chart interpretations which he assembled with the help of his staff and sent out in the mail. His was the first modern astrological business.

Alan Leo has also been credited with the creation – or at least the popularization – of sun-sign astrology. Before Alan Leo, the Ascendant was the most important feature in the birth chart. But according to the esoteric tradition, the sun is the spiritual center of the universe. Leo was born William Frederick Allan but changed his name to Alan Leo in honor of his sun which, since he was born August 7, was in the sign of Leo. The sun-sign astrology columns in newspapers and magazines with which everyone is familiar are part of the legacy of Alan Leo.

The planet (or dwarf planet, according to the International Astronomical Union) Pluto was discovered in 1930 by Clyde Tombaugh as the result of research begun in 1905 by Percival Lowell. If Uranus is the planet of the eighteenth century and Neptune is the planet of the nineteenth century, Pluto is – for astrologers, anyway – quintessentially the planet of the twentieth century. According to *Larousse*, cultural trends appearing around the time of its discovery include:

the rise of Nazism and fascism in Europe; the discovery of atomic energy; the laboratory perfection of television equipment and the rise of mass media; the growing public acceptance of psychoanalysis; and the sudden prevalence of cancer.

Astrologers are beginning to associate Pluto with a subtle but powerful underground force that lies dormant within various collective systems and bursts forth volcanically at a given moment: the power of the masses and the collective unconscious; the invisible power of the atom; the power of the unconscious, forcing to the surface repressed memories locked within the cells; the dark, proliferating force of cancer that works insidiously within the body until it is diagnosed. Pluto's qualities are power, elimination, latency, eruption, annihilation, transformation, renewal, regeneration.[155]

The great Swiss psychologist Carl Gustav Jung (1875-1961) was a lifelong believer in and advocate for astrology. Jung saw the planets as archetypes: universal forms somewhat similar to Platonic ideas. He began studying astrology in 1910, just three years before he broke with Freud. Jung believed that astrology represented the sum of all the psychological knowledge of antiquity.[156] He famously wrote: "Whatever is born or done at this particular moment of time has the quality of this moment of time."[157]

In a letter to the Indian astrologer B. V. Raman, he wrote:

As a psychologist, I am chiefly interested in the particular light the horoscope sheds on certain complications in the character. In cases of difficult psychological diagnosis I usually get a horoscope in order to have a further point of view from an entirely different angle. I must say that I very often found that the astrological data elucidated certain points which I otherwise would have been unable to understand.[158]

On the question of how astrology works Jung alternated between the notion of some type of causality or energy not yet understood, and his theory of synchronicity, first formulated in the 1920s. He defined synchronicity as "the simultaneous occurrence of two meaningfully, but not causally, connected events," or more simply, "an acausal connecting principle" illustrating the possibility of "meaningful coincidence."[159]

Jung's idea of "meaningful coincidence" has been dismissed as a contradiction in terms, but this argument works only if one ignores the original meaning of "coincidence," i.e., "occupying the same period of space or time," (or as Jung put it, "a falling together in time")[160] and embraces only the later connotation of randomness.

Jung wanted astrology to become a science, and he thought that statistics should be used to establish a scientific basis for astrology. In his essay "Synchronicity: An Acausal Connecting Principle," he discusses an experiment he did with hundreds of birth charts of married couples. Although he concedes, with characteristic humility, that his research was flawed in several ways, he did come up with results that supported astrological principles about the importance of Sun, Moon, and Ascendant in contacts between charts.[161]

In this essay, Jung refers to the experiments in ESP of J. B. Rhine as a possible example of synchronicity. He found Rhine's results an intriguing challenge to the materialist paradigm.

The Myers-Briggs Type Indicator, a personality test that has become a standard tool of modern psychology, was extrapolated from Jung's theories in his 1921 book *Psychological Types*. Jung's four types – sensation, intuition, thinking, and feeling – were in turn based on the four elements of classical cosmology: earth, fire, air, and water, and the corresponding temperaments or humors: melancholic, choleric, sanguine, and phlegmatic. Campion sees the Myers-Briggs test as an example of "disguised astrology."[162] He writes that Jung gave astrology "an intelligent modern voice, allowing it to appeal to a much wider educated constituency than had previously been the case."[163]

Patrick Curry, in his *Prophecy and Power: Astrology in Early Modern England*, (1989, Oxford: Polity Press), adopts a class model to explain astrology's decline and eventual revival. Corresponding to the upper, middle, and lower classes of society, he uses a classification of high, middling, and low astrology. Great intellects like Kepler and Francis Bacon practiced "high" astrology. Professionals like William Lilly who made their living casting horoscopes belonged to the "middle" class of astrologers. The authors of the almanacs and the country folk who relied on them were examples of "low" astrology. There was a certain amount of overlap in these categories (i.e., Kepler cast horoscopes for money and Lilly published almanacs).

Charles Carter (1887-1968), distinguished English astrologer and author, wrote that

Jung's "open advocacy of Astrology has probably done much to render the subject respectable to our day."[164] While this is a bit of an overstatement – astrology has yet to become respectable – it is certainly true that over the course of the twentieth century astrology began to engage the attention of educated persons, for many of whom Jung was a source of inspiration.

The shining example of this new breed of modern astrologers is the American astrologer, author, and composer Dane Rudhyar (1895-1985), who founded humanistic astrology. A Theosophist – he was Alice Bailey's most distinguished student – Rudhyar combined a solid grounding in oriental philosophy with a deep immersion in the ideas of Jung.

Writing in the 1930s, he saw traditional European astrology with its emphasis on the prediction of events as spiritually bankrupt. In the most famous of his thirty books, *The Astrology of Personality* (1936), he called for a reformulation of astrological principles in the light of contemporary depth psychology. He believed that astrology should concern itself primarily with people rather than with events, which is why he called his approach "humanistic." Rudhyar was not interested in astrology as an empirical science or the planets as transmitters of influence; he saw astrology as a symbolic language, like mathematics, and the planets as symbols of personality functions and timers of cycles. The birth chart was a map of the psyche.

Rudhyar's impact on modern astrology has been tremendous. Campion writes that

somewhere around ninety percent of modern astrologers . . . think that the universe speaks in metaphors, agreeing with the definition of astrology as a language . . . that the birth chart represents a map of our potential, which it is up to each individual to develop as best they can.[165]

At the same time that Rudhyar was beginning to write, eloquently, his works of "high" astrology, the kind of "low" or popular astrology that had survived in the almanacs was enjoying a new incarnation in the sun-sign columns of newspapers and magazines. In England, the first regular newspaper column devoted to astrology appeared in the British *Sunday Express* in 1930 and featured an analysis of the horoscope of the infant Princess Margaret Rose. The first sun-sign column, "Your Stars," by British astrologer R. H. Naylor, appeared in *Prediction* magazine in

1936. In America, the first sun-sign columns probably appeared around 1930.

Dane Rudhyar actually encouraged the sun-sign columns because he felt they would help to spread astrology. He even wrote columns himself for the magazine *American Astrology*. Other distinguished American astrologers have lent their talents to these newspaper and magazine columns, including the late great Charles Jayne (1911-1985). The nuggets offered for each of the twelve sun signs have tended to focus on advice about how to make the most of the planetary energies rather than prediction of events over which the reader has no control.

When I began studying astrology in the late sixties, the books of Dane Rudhyar and Marc Edmund Jones (1888-1980) were a breath of fresh air among the offerings that were then available. Rudhyar's little book *The Practice of Astrology* made a great impression on me because of its emphasis on the importance of ethics and the responsibility of the astrologer, especially with regard to prediction:

Prediction has value only as it contributes to the person's development and essential welfare. . . . Fortune-telling is an unorganized attempt at haphazard prediction on the basis of isolated and incomplete data. Its purpose is at best to satisfy the apparent curiosity of the client; at its worst, to pander to his insecurity and his fears for the sake of profit. . ..

Above all, [the astrologer] should understand that the value of astrology – psychologically speaking, at least – does not reside in his ability to tell what is *likely* to happen (he can never say more!) as much as to help the client to understand fully and in terms of his total being what is happening, or has already happened.[166]

One result of the decline of astrology in the seventeenth and eighteenth centuries was the relative isolation of astrologers, the lack of support from a community of their colleagues. With the revival of high astrology in the late nineteenth and early twentieth centuries, this situation began to change. Astrologers started to organize, to form professional organizations designed to teach astrology, do research, and establish professional standards.

The American Federation of Astrologers (AFA) was incorporated in 1938 in Washington, D.C. "to encourage and promote the science and art of astrology through research, teaching, lecturing and practice . . . and to establish as high a standard of professional ethics for legitimate astrologers as now exists in other educational professions. . ..." AFA's examination and certification program has been in existence since 1960.[167]

In England in 1948 the Astrological Lodge of the Theosophical Society set up a school called The Faculty of Astrological Studies, and in 1958 in London, a group of Lodge members formed the Astrological Association of Great Britain (AA). Its aims and objectives are "to promote scientific research and education and to work for a more widespread understanding of astrology as an art and science." It publishes a number of scholarly journals.[168]

Founded in 1968, the International Society for Astrological Research (ISAR) offers "a certification program that emphasizes ethics awareness, competence in techniques and interpretation, and consulting skills." ISAR networks with other organizations, sponsors research, and provides a professional journal and a weekly e-newsletter. The first international astrological organization, ISAR has members in 49 countries.[169]

The National Council for Geocosmic Research (NCGR) was formed in 1971 in Massachusetts. The term "geocosmic" is defined as "of or pertaining to the study of correspondences and cycles involving earthly phenomena and cosmic (celestial) events." NCGR is dedicated to raising the standards of astrological education and research. Though it began as "national," its current membership includes a growing number of international members.

NCGR administers a four-level education and testing program leading toward certification for astrologers. A code of ethics, to which members are accountable, provides guidance for the practice of astrology. NCGR sponsors an educational conference every two or three years at various venues around the U.S. and publishes a biannual journal and a newsletter.[170]

The Association for Astrological Networking (AFAN) was formed in 1983 as a reform movement within the AFA. It was galvanized by the arrest in San Jose, California of Shirley Sunderbruch, a fifty-six-year-old astrologer who was handcuffed by the police in a sting operation and evicted from her retirement community. Her astrological books, tape recorder, and prepared readings were confiscated and Sunderbruch, a heart patient, barely escaped being jailed. She was finally exonerated two years later through the efforts of AFAN's legal committee. In over thirty years of effort, AFAN has never lost a challenge against discriminatory anti-astrology laws.[171]

There are still anti-astrology statutes in many cities and states, and local police and religious fundamentalists who are eager to criminalize astrology. An important project of these astrological organizations is media watch: keeping track of how astrology and astrologers are portrayed in the media, for it is public attitudes toward astrology that will ultimately determine its legality.

On July 4, 2015, I counted seventeen international astrological organizations on the Web and too many local ones to count.

Kepler College was incorporated in 1992 in Seattle as an institution of higher learning approved by the state of Washington that focused on interdisciplinary liberal arts with an emphasis on the history of astrology. The mathematics and practice of astrology are taught, using techniques ranging from the Vedic to Western systems and from the Hellenistic period to the present.

After the college opened its doors in 2000, Associate's, Bachelor's and Master's of Arts degrees were authorized by the state's Higher Education Coordinating Board, but this authorization did not constitute endorsement. Indeed, the school has been criticized for teaching astrology and the Board has been criticized for allowing it to do so. John Silber, Chancellor of Boston University, has said that the college's choice of name "honored Kepler not for his strength but for his weakness." He went on,

> The fact is that astrology, whether judged by its theory or its practice, is bunkum. In a free society there is no reason to prevent those who wish to learn nonsense from finding teachers who want to make money peddling nonsense. But it is inexcusable for the government to certify teachers of nonsense as competent or to authorize – that is, endorse – the granting of degrees in nonsense.[172]

John Silber belongs to the long line of mainstream academics who prefer to believe that Kepler had no respect for astrology and only practiced it for the money, whereas an unbiased study of history shows that Kepler never stopped believing in what he saw as the essential core of astrology, and energetically advocated its reform.

In 2010 the economic recession forced Kepler College to abandon its efforts to receive accreditation from the state and thus to discontinue its program as a degree-granting institution. Since then, Kepler College has offered professional-quality education for those interested in astrology in the form of workshops and online courses leading to a certificate of completion.[173]

This summary of twentieth-century astrology would be incomplete without mention of the extraordinary body of work of astrologer, teacher, and translator Robert Hand. Holder of a B.A. in history from Brandeis and an M.A. in history from the Catholic University in Washington, D.C., Hand has done graduate work at Princeton, and wrote his dissertation for the Catholic University's Ph.D. on Medieval and Byzantine Studies.

In addition to writing seven classic astrology texts, Hand has founded a number of astrology-related companies. He began writing computer programs in 1977 in order to bring the benefits of fast and accurate calculation to the practice of astrology. In 1980 he founded Astro-Graphics Services, a provider of astrological software, which later became Astrolabe.

In 1993, together with Robert Schmidt, Ellen Black, and Robert Zoller, Hand co-founded Project Hindsight in order to make primary source texts in Greek, Latin, Hebrew, and Arabic available in modern English translations. And in 1997 Hand founded the Archive for the Retrieval of Historical Astrological Texts. ARHAT is a formal archive, library and publishing company which specializes in the translation of ancient, classic, and medieval texts previously unavailable in English. In 2008, at the United Astrology Conference (UAC) in Denver, Hand received the Regulus Lifetime Achievement Award for his many decades of service to astrology.

The earliest computer-generated horoscopes were available from Astroflash, an enormous IBM computer in a small fiberglass cubicle that appeared in New York's Grand Central Station in 1969. The price of a combined portrait and six-month forecast – fourteen connected pages with punch holes on either side – was five dollars. Poet Archibald Macleish, writing about Astroflash in the *Harvard Crimson* that year, described himself as "fascinated," and admitted that "the machine was right, upsettingly accurate, again and again."[174]

The impact of high-speed personal computers on astrology has been enormous. In the first place, the computer programs now available have greatly simplified the astrologer's work, especially for beginners: no more juggling of slide rules, ephemerides, and tables of houses. What took me twenty minutes in the sixties now takes a few seconds at the keyboard.

More importantly, the sophisticated software developed by pioneers like Michael Erlewine and the late Neil Michelsen has given modern astrology a tremendous leap forward in terms of mathematical accuracy and the capacity for statistical research involving large samples.

The work of Michel and Françoise Gauquelin is unquestionably the most important astrological research of the twentieth century. Over the course of their careers the Gauquelins collected over 20,000 dates of birth of eminent professionals. When Michel Gauquelin began collecting birth data in 1949, all of his calculations had to be done by hand. In the late 1970s in San Diego, Neil Michelsen placed his early calculation programs at the disposal of the Gauquelins, enabling them to streamline their research.

When I wrote about Gauquelin in my column in the *Woodstock Times* back in the seventies, I said that he was "initially a skeptic." I was mistaken. I did not realize that Gauquelin's fascination – obsession, really – with astrology dated from early childhood. His father, a dentist, taught him to calculate birth charts, and his schoolmates gave him the nickname "Nostradamus" because of his early proficiency in that art.

But Gauquelin wanted to make astrology scientific, and he began his research with an eye to eliminating anything that did not stand up to statistical testing. Like Kepler, he threw out almost all of traditional astrology: the signs of the zodiac, the houses, transits, directions – even half of the planets. He eliminated the Sun, Mercury, and the three modern planets, Uranus, Neptune, and Pluto, retaining only the positions of the Moon, Venus, Mars, Jupiter, and Saturn relative to the Earth's daily motion. Like Kepler, he was so critical of contemporary astrology that he was often assumed to be anti-astrology, whereas in fact he was seeking that "grain of gold" that he was convinced was embedded somewhere in the mass of superstition.

The Gauquelins' research program, which extended over more than forty years, focused on three main areas of study: the positions of four planets and the Moon in the birth charts of eminent professionals; the positions of those bodies in the charts of parents and children, what Gauquelin called "planetary heredity;" and their positions in the charts of persons with the character traits associated with certain professions. Although the last two areas produced positive results, it was the work on the birth charts of successful professionals that had the most dramatic results. And of these, it was one isolated result – the mysterious predictability of the positions of the planet Mars in the charts of sports champions, the so-called "Mars effect" – that provoked a controversy that went on for over forty years.

To state it as simply as possible, Gauquelin (and after his wife, Françoise, joined him, the Gauquelins) found that in samples of at least 300 birth charts of sports champions the planet Mars was found either rising or culminating with a frequency that exceeded chance, chance being determined by a much larger sample of the charts of ordinary persons. Champions were distinguished from lesser athletes by setting records or by being chosen for major competitions. The criterion of eminence turned out to be critical, as the Mars effect disappeared for athletes not classified as eminent.[175]

Obviously, any solid evidence of an effect of the planets on human behavior presents scientists with an almost intolerable challenge. Astronomer George Abell wrote in 1979, "If the Gauquelin planetary effects are real, they are truly astonishing and lie beyond anything that science can at present understand."[176]

A major tenet – if not the very foundation – of the scientific method is that the results of research, to be accepted as sound, must be capable of replication. There were three major attempts by skeptical groups to replicate (more accurately, to disprove) the Gauquelin results, one in the U.S. and two in Europe.

In 1956, the *Comité Belge pour l'Investigation Scientifique des Phénonèmes Réputés Paranormaux* (Belgian Committee for the Scientific Investigation of So-called Paranormal Phenomena), *Comité Para* for short, founded in 1948, initially refused Gauquelin's request to test his results, claiming that "professional astronomers have studied the question *a priori*. . .." Eleven years later, they finally agreed to run their own test with 535 new sports champions. When their results strongly supported Gauquelin's findings they refused to publish them for another eight years, and when finally forced to do so in 1976, they accused Gauquelin of imaginary demographic errors.

In 1976, the American Committee for the Scientific Investigation of Claims of the Paranormal (CSICOP) was founded by Paul Kurtz, a professor of philosophy who was also editor of *The Humanist*, the magazine that had published an "Objections to Astrology" statement signed by 186 scientists. That issue of the magazine had criticized the Gauquelins' work and the Gauquelins had objected. The result was that Marvin Zelen, a professor of statistics at Harvard and a member of CSICOP, performed a test that clearly indicated that the Gauquelins were correct and that the "demographic errors" cited by the *Comité Para* were imaginary. However, the article reporting the results in *The Humanist* violated elementary statistics in an effort to disprove the Mars effect, even though the editors were privately aware that Gauquelin's calculations were correct. The whole shameful story has been documented several times, first by astronomer and former CSICOP member Dennis Rawlins in *Fate* magazine, and later by academics Patrick Curry, John Anthony West, and Kenneth Irving.[177]

In the meantime, Paul Kurtz undertook to replicate the Mars effect using a group of American athletes. (One reason for the focus on the planet Mars and sports champions rather than the other professional studies is the fast turnover of eminent athletes, making fresh data more easily available.) Kurtz's method of data collection lacked the scrupulous controls and transparency of the Gauquelins' work, and his sampling procedures violated the all-important criterion of eminence. A report by Kurtz, Zelen, and Abell on the CSICOP's study of American athletes was published in *The Skeptical Inquirer*. After Dennis Rawlins' exposé appeared in 1981 several members of CSICOP quit the organization, but the Mars effect controversy remained unresolved.

In France, most scientists had studiously ignored the work of their compatriot, but in 1982 a third group of skeptics, the *Comité Français pour l'Etude des Phénonèmes Paranormaux* (French Committee for the Study of Paranormal Phenomena), CFEPP for short, got into the act. These gentlemen ignored the eminence criterion and violated their own protocol, which specified that they stay in touch with Monsieur Gauquelin and that no partial results of the experiment be published until their complete and final report. CFEPP failed to communicate with Gauquelin for eight years and leaked false conclusions to the press. These appeared under the headlines "Mars leaves astrologers in the lurch" and "Astrology's last holy cow killed."[178] As of 1996, the French committee had still not published their results.

In 1985, while the French committee was incommunicado, Suitbert Ertel, a German professor of psychology at the University of Göttingen, decided to undertake his own study of the Mars effect. Ertel understood that the criterion of eminence was crucial to the debate and that what was missing in the previous research was an objective system for determining eminence. The method he proposed ranked sports champions by the number of citations in a fixed set of eighteen reference books. He carried out his investigation in Gauquelin's Paris laboratory, where he was allowed full access to all his data. His results vindicated the Mars effect and showed that the failure of the CSICOP study to replicate Gauquelin's findings was due to violation of the eminence requirement.[179]

In 1993-94, Arno Müller and Suitbert Ertel replicated Gauquelin's findings on the positions of Mars and Saturn in the charts of physicians. They obtained essentially the same results as he did, indicating that "his data collection was unbiased, his methodology was correct, and his conclusions were sound."[180]

But in 1991 Gauquelin, who had fallen into a depression, had died by his own hand.

In a tribute to Michel Gauquelin in the British *Astrological Journal*, Françoise Schneider-Gauquelin wrote:

I think that astrology needs the kind of serious sorting out of what has a lasting scientific meaning among the innumerable ideas and techniques offered to public scrutiny in astrological journals. . . .. Not only was he one of the most dedicated and scrupulous collectors of data in the world; he had also the grand vision of how many thoroughly researched data samples, replications, and controls were needed to sound reasonably convincing in scientific spheres. Obviously, after so many battles successfully conducted against rivals who were not always fighting in as fair and objective way as he did, weariness overwhelmed him.[181]

After his death, all of Gauquelin's files were reported to have been destroyed and his library sold by his heirs.[182]

The astrological community was shocked and saddened by the death of this indefatigable researcher, and even some scientists paid tribute to his work. Back in 1982, British psychologist Hans Eysenck of the London University of Psychiatry had written, "Emotionally, I would prefer that Gauquelin's results don't hold, rationally I must accept that they do."[183] Geoffrey Dean, a fellow of the Committee for Skeptical Inquiry, wrote, "From his tiny Laboratoire in the backstreets of Paris, his immense labours created mesmerising puzzles that have assured his place in history."[184] And in their book *Astrology: Science or Superstition?*, after a careful examination of Gauquelin's work, Hans J. Eysenck and David Nias had concluded:

We feel obliged to admit that there is something here that requires explanation. However much it may go against the grain, other scientists who take the trouble to examine the evidence may eventually be forced to a similar conclusion. The findings are inexplicable but they are also factual, and as such can no longer be ignored: they cannot just be wished away because they are unpalatable or not in accord with the laws of present-day science. . .. Perhaps the time has come to state quite unequivocally that a new science is in process of being born.[185]

Although he regarded much of modern astrology as without scientific foundation, Gauquelin did provide solid statistical evidence, as Kenneth Irving has noted, for at least three of its fundamental principles: the centrality of the planets; the distinctive character of the planets; and the importance of angularity. Irving writes: "Whatever the faults of modern astrology may be, one can see from the Gauquelin planetary effects that it is based, at least in part, on observation rather than magic or myth."[186]

In the last of his twenty books, published after his death, Gauquelin wrote:

On the whole, it emerged that there was an increasingly solid statistical link between the time of birth of great men and their occupational success. . .. Having collected over 20,000 dates of birth of professional celebrities from various European countries and from the United States, I had to draw the unavoidable conclusion that the position of the planets at birth is linked to one's destiny. What a challenge to the rational mind![187]

And back in 1983, he had written:

After thirty years of critical consideration of astrology, my passion for it has not diminished. But today I would not allow myself to draw drastic conclusions as I have sometimes done in the past. I will be content simply to have thrown a little light on this vast mystery which has occupied so many great minds over the centuries.[188]

Under the auspices of the Urania Trust, a British non-profit organization dedicated to the study of astrology, the Michel Gauquelin Research Fund has been established, with pledged backing from both the Astrological Association and the Faculty of Astrological Studies. The purpose of this fund is to encourage astrological research at the highest level; in essence, to continue the work of Michel Gauquelin.[189]

The Gauquelins' forty years of research addresses what Kenneth Irving calls the "metric" layer of the horoscope, which is purely astronomical, as it consists mainly of the celestial sphere and the planets. He distinguishes this from the "mantic" (prophetic or divinatory) layer, which consists of the zodiac, the houses, and the rest of traditional "judicial" astrology. He writes:

Over many decades of astrological research, studies that engage the mantic layer have produced little of substance. [The work of the American psychologist Vernon Clark (1911-1967) belongs to this group of projects.] On the other hand, studies aimed at the metric layer of the horoscope have produced clear evidence for the interaction of the cosmos with human beings.[190]

For the most part, mainstream scientists are unimpressed. Evolutionary biologist Richard Dawkins, Professor of the Public Understanding of Science at Oxford University and a CSICOP fellow, advocates the prosecution of astrologers for fraud.[191]

On the other hand, American astronomer and author Carl Sagan (1934-1996), who had no love for astrology, declined to sign the *Humanist* anti-astrology manifesto of 1975. In a letter to the editor, he wrote:

I find myself unable to endorse the "objections to astrology" statement, not because I feel that astrology has any validity whatever, but because I felt and still feel that the tone of the statement is authoritarian. The fundamental point is not that the origins of astrology are shrouded in superstition. This is true as well for chemistry, medicine and astronomy, to mention only three.

To discuss the psychological motivation of those who believe in astrology seems to me quite peripheral to the issue of its validity. That we can think of no mechanism for astrology is relevant but unconvincing. No mechanism was known, for example, for continental drift when it was proposed by [Alfred] Wegner. Nevertheless, we see that Wegner was right, and those who objected on the grounds of an unavailable mechanism were wrong.[192]

Campion notes that critics of the *Humanist* statement

have identified a moral anger in the statement, driven by an unsupported assumption that astrologers cannot genuinely believe in something they know to be false and are therefore charlatans, defrauding a public which is in need of protection.[193]

Campion puts his finger on the deeper reason for the opposition to astrology on the part of science:

Astrology is subject to public ridicule and condemnation precisely because its world-view challenges the scientific and religious orthodoxy that attempts to stamp out other world views. Whereas science tends to regard the universe as devoid of meaning and conventional religion focuses meaning on God, astrology maintains the ancient idea that the universe is in itself meaningful, that meaning and purpose are woven into its fabric.[194]

(9) The Twenty-first Century

While the view that all is accidental and random and that the universe consists of dead matter without purpose or meaning dominated modern science in the twentieth century, there are, at the dawn of the twenty-first century, clear indications that that view is being questioned and that the deeply entrenched prejudice against astrology is at least being reconsidered.

Sometime before his death in 1961 Jung, who referred to astrology as "*scientia intuitiva*," wrote that astrology was "knocking at the doors of the universities from which it was banished some three hundred years ago."[195] While the doors of the universities have not been flung open across the land, it is certainly true that astrology is once again engaging the attention of educated persons with first-class minds.

The publication of Nicholas Campion's two-volume history of astrology in 2008 and

2009 is in itself an event in the history of astrology. At last we have a history of astrology that is neither hostile nor tendentious but open-minded, nuanced, and comprehensive. Campion wrote his Ph.D. thesis at the University of the West of England in 2004 on "Prophecy, Cosmology, and the New Age Movement: The Extent and Nature of Contemporary Belief in Astrology."

And at Indiana University in 2002, Darrel Rutkin wrote his Ph.D. thesis on "Astrology, Natural Philosophy, and the History of Science, C. 1250-1700: Studies toward an Interpretation of Giovanni Pico della Mirandola's *Disputati*." There's a whole new generation of scholars who are open to astrology and of astrologers with advanced degrees. When I asked Darrel Rutkin whether he "believes" in astrology his answer was, "I don't believe in it – but it works!"

Another academic who has turned his attention to astrology is Richard Tarnas, graduate of Harvard University and Saybrook Institute and professor of philosophy and cultural history at the California Institute of Integral Studies in San Francisco. Author of *The Passion of the Western Mind: Understanding the Ideas That Have Shaped Our World View* (1991), which Joseph Campbell called "The most lucid and precise presentation I have read of Western thought," Tarnas more recently published *Cosmos and Psyche: Intimations of a New World View* (2007), in which he presents thirty years of meticulous historical research in astrology.

Initially skeptical about astrology, which he describes as "one special, highly controversial class of synchronicities," Tarnas gradually became impressed by the massive number of precise correspondences he discovered between planetary configurations and important events in human history. For example, here is Tarnas on the timing of the Uranus opposition (the transit that occurs in everyone's life in their early forties) in the biographies of some key cultural figures:

I discovered that when Galileo made his first telescopic discoveries between October 1609 and March 1610 and then quickly wrote and published *Sidereus Nuncius* ("The Starry Messenger"), which heralded the truth of the Copernican theory and caused a sensation in European intellectual circles, he had the identical personal Uranus transit that René Descartes had in 1637 when he published his equally epoch-making *Discourse on Method*, the manifesto of modern reason and the foundational work of modern philosophy. Moreover, this also happened to be the same transit Isaac Newton had in 1687 when he published the *Principia*, the foundational work of modern science. [196]

The bulk of Tarnas' book is devoted to mundane astrology, the branch of astrology that deals with world events and cultural trends rather than with the birth charts of individuals. In her book *Astrology: Its History and Influence in the Western World* (1942), American astrologer Ellen McCaffery (1886-1953) wrote that it is easier to demonstrate the validity of astrology through mundane rather than natal astrology because people as a group have less control over their destiny.[197] Be that as it may, Tarnas' studies of the correlations between the major cycles of the outer planets and pivotal events in world history are just as compelling as the natal material in the work of Gauquelin and others.

Mundane astrology is especially concerned with the patterns formed in the sky by the outer planets – Jupiter, Saturn, Uranus, Neptune, and Pluto – with one another rather than with the patterns formed by the planets to the charts of individuals. Just as our Sun and Moon have a monthly cycle from New Moon to Full Moon and back to New as the Moon forms a conjunction to the Sun, waxes to the opposition, and wanes back to the next conjunction, so each pair of outer planets has a similar cycle of conjunctions and oppositions over the course of years.

For example, the Jupiter-Saturn cycle, which has been correlated with the deaths of presidents in American history, has an interval between conjunctions of a little under twenty years. But the synodic cycles of the slower moving modern planets – Uranus, Neptune, and Pluto – have much longer intervals. The conjunctions and oppositions are relatively rare occurrences, but when they do occur, they can be in orb for as long as a decade.

Tarnas found that when he applied the meanings that have been assigned to the modern planets to periods of history when those planets were in major alignment (conjunction or opposition), he was impressed by the correlations that emerged:

These extended alignments of the outer planets consistently seemed to coincide with sustained periods during which a particular archetypal complex was conspicuously dominant in the collective psyche, defining the zeitgeist, as it were, of that cultural moment. . ..

One of the first such instances was the decade of the 1960s. By all accounts the Sixties were an extraordinary era. . . . The entire decade seems to have been animated by a peculiarly vivid and compelling spirit – something "in the air" – an elemental force apparent to all at the time, that was not present in such a tangible manner during the immediate or subsequent decades, and that in retrospect still sets the era apart as a phenomenon unique in recent memory. Early in the course of my research I noticed that during the entire period of this decade, specifically from 1960 to 1972, there took place a conjunction of two outer planets, Uranus and Pluto, that occurs relatively rarely. Indeed, this was the only conjunction of these planets in the entire twentieth century.[198]

Tarnas noticed that when two outer planets were aligned, their assigned archetypal meanings seemed to combine. In the case of Uranus and Pluto, the rebellious, liberating, innovative, and disruptive energy of Uranus was combined with the intense, powerful, compelling, and transformative energy of Pluto. He found to his surprise that the same two bodies were aligned, this time in an opposition, during the entire decade of the French revolution. Between the French revolution and the 1960s there were only two other periods when these two planets were aligned, and both were periods of massive change, innovation, and upheaval. The first was the decade between 1845 and 1856, a time of radical political and social movements which saw the revolutionary seeds planted by Marx and Engels, Thoreau, Frederick Douglass and Harriet Tubman, and Elizabeth Cady Stanton and Susan B. Anthony. The second alignment of Uranus and Pluto occurred between 1896 and 1907, which saw the rise of socialism, communism, and the labor movements, women's suffrage, and civil rights, not to mention the breakthroughs in the work of Einstein and Freud.

Tarnas goes on to explore the cycles of the outer planets in relation to landmarks in science and religion, art, music, and literature, politics and war. Like astrological reformers Kepler and Gauquelin before him, Tarnas ignores the signs of the zodiac (thus bypassing the tropical-sidereal issue) and focuses his attention on the planets. And like them, his research presents compelling evidence of correlations between the motions of the planets and events on Earth: evidence of a core of validity in the astrological hypothesis that our modern science cannot explain.

Another important book which, as I recall, never mentions the word astrology but sheds light on the current divide between astrology and science is Iain McGilchrist's *The Master and his Emissary: The Divided Brain and the Making of the Modern World.* A former Oxford literary scholar turned physician and psychiatrist, McGilchrist draws upon his work as a clinical psychologist, his neuroimaging research at Johns Hopkins University in Baltimore, and his familiarity with the work of other neurologists to arrive at a unique perspective on the functioning of the human brain. Although he finds the traditional model of the division of labor between the two hemispheres – reason in the left brain, creativity in the right – simplistic, and maintains that normally the two hemispheres should work together, he does find that each hemisphere has a different view of the world.

The right hemisphere – the Master in his analogy – has grounding and integrating functions. It is good at holistic or gestalt perception and pattern recognition, and has an intuitive understanding of metaphor, humor, and music. The left hemisphere – the Emissary – is good at taking things apart, at abstraction (which literally means removing things from their context) and is at home with tools, concepts, and words. Both are necessary, but things go better when the right hemisphere, with its broader range of attention, is in charge. Complex symbols, such as archetypes, belong to the right hemisphere. Simple symbols, like a traffic light, belong to the left hemisphere. Machines: left brain. Living beings: right brain. The essential difference in the way the two hemispheres work lies in the nature of attention:

Different aspects of the world come into being through the interaction of our brains with whatever it is that exists apart from ourselves, and precisely *which* aspects come into being depends on the nature of our attention. . .. Attention is a moral act: it creates, brings aspects of things into being, but in doing so makes others recede.[199]

The left hemisphere favors a narrow window of attention. Yet, it has no monopoly on reason or language. Some types of reasoning – e.g., deduction – are a right brain activity. Aha! moments, insight, higher math solutions all belong to the right hemisphere, along with Pythagoras and the Music of the Spheres.

The left hemisphere tends to insist on its theory at the expense of getting things wrong, while the right hemisphere has a tolerance for uncertainty. It has, if you will, Keats' negative capability: the capacity to "be in uncertainties, mysteries, doubts, without any irritable reaching after fact and reason."

In the right hemisphere, opposites are not incompatible. The right hemisphere understands Jung's *coniunctio oppositorum*, the union of opposites; the left hemisphere subscribes to the law of the excluded middle: a proposition is either true or false; no resolution is possible.[200]

It seems clear that astrology is a right-brain activity. Planets in opposition have much in common and the energies they represent have a potential for integration. Astrology is all about the big picture, and the planetary archetypes are a form of metaphor. According to McGilchrist, metaphor underlies all forms of understanding, science and philosophy no less than poetry and art.[201]

In short, the two hemispheres have two points of view, two ways of being in the world. The right hemisphere is self-correcting, the left hemisphere is self-perpetuating. Much of modern science is a left-brain activity. We need to be able to make fine discriminations and to reason, but only in service to the wholeness that the right hemisphere can bring.

Like Richard Tarnas, Iain McGilchrist brings a profound knowledge of Western history to bear on his ideas. And he concludes that whereas in periods of renaissance the two hemispheres have been largely in balance, with the rise of scientific materialism and bureaucracy over recent centuries there has been a relentless erosion of the power of the right hemisphere, with unfortunate – and even dangerous – results:

Today all the available sources of intuitive life – cultural tradition, the natural world, the body, religion and art – have been so conceptualised, devitalised and "deconstructed" . . . by the world of words, mechanistic systems and theories constituted by the left hemisphere that their power to help us see beyond the hermetic world that it has set up has been largely drained from them. . . . A number of influential figures in the history of ideas, among them Nietzsche, Freud and Heidegger, have noted a gradual encroachment over time of rationality on the natural territory of intuition or instinct.[202]

This encroachment of rationality over intuition – this usurpation of power by the left hemisphere – has helped to produce what Max Weber[203] called the "disenchanted" world, a world devoid of meaning and purpose, a world consisting of dead matter where we have succeeded, as Heidegger put it, in turning nature into one gigantic filling station.

McGilchrist blames the Reformation and the Enlightenment, both left hemisphere movements that sought to replace the image with the word. He sees the Reformation as a failure of metaphor, and he blames the Enlightenment for the doctrine of the infallibility of science, which has become the dominant myth of the twentieth century.

Einstein is often quoted as having said, "The intuitive mind is a sacred gift and the rational mind is a faithful servant. We have created a society that honors the servant and has forgotten the gift." Whether or not Einstein actually said this, he certainly understood that reductionism could only succeed within a self-enclosed system:

> . . . the supreme task of the physicist is the discovery of the most general elementary laws from which the world-picture can be deduced logically. But there is no logical way to the discovery of these elemental laws. There is only the way of intuition, which is helped by a feeling for the order lying behind the appearance, and this *Einfühlung* [literally, empathy, or "feeling one's way in"] is developed by experience.[204]

The left brain's failure to understand metaphor – and its dismissal of everything it fails to understand as "magical thinking" – may account at least partly for the attitude of modern science toward astrology. The narrow window of attention favored by the left brain has brought about a failure of the imagination, a failure to see that imagination – intuition, inspiration, dreams, and even visions – have a role to play in science. And its need for certainty tends to close the window of its attention with a bang. McGilchrist notes that devout believers in the random universe subscribe to their own form of fundamentalism:

> In the field of religion there are dogmatists of no-faith as there are of faith, and both seem to me closer to one another than those who try to keep the door open to the possibility of something beyond the customary ways in which we think. . ... Certainty is the greatest of all illusions: whatever kind of fundamentalism it may underwrite, that of religion or that of science, it is what the ancients meant by *hubris*.[205]

Lately I've noticed a few indications of a possible undermining of the left brain's clinging to its world view and specifically of its rational materialist dismissal of astrology.

Dava Sobel, former *New York Times* science reporter and winner of many awards for her contributions to the public understanding of science, may have had what Jung called "an essential change of attitude." In her 1999 book *Galileo's Daughter*, a finalist for the 2000 Pulitzer Prize in biography, she makes no mention of Galileo's astrology. But in her 2005 book *The Planets*, in her chapter on Jupiter, we come upon this astonishing paragraph:

Had astronomy and astrology not parted ways so long ago, some of the *Galileo* mission's problems might have been foreseen. A natal chart drawn for *Galileo*, "born" at Cape Canaveral on the day of its launch, October 18, 1989, reflects a strong, even aggressive spacecraft, with the Sun in Libra for balance, and Mars conjunct with the Sun at the midheaven, adding ambition. At the ascendant, Saturn, Uranus, and Neptune cluster together, which lends a sense of seriousness and importance to the venture. Mercury, however, the planet of communication, makes the worst possible angle – a square, or negative aspect – with Jupiter's position. Another unfortunate Mercury square opposes the powerful triad of Saturn, Uranus, and Neptune. The chart shows Jupiter occupying *Galileo*'s seventh house, the mansion of marriage and partnership. Surely the spacecraft partnered with Jupiter through its life work, and also united with Jupiter in its ultimate fate.[206]

All this is presented seamlessly along with the astronomical data, with no apology and indeed no warning that we are about to venture into the forbidden territory of astrology.

Inevitably, *The Planets* received mixed reviews. William Grimes at *The New York Times* ignores the astrology altogether. David Perlman at the *San Francisco Chronicle* complains that there's "too much mixing of astrology and astronomy," but forgives her because "she is, after all, an expert on Galileo, and he was no mean disciple of astrology in his day." John Morrish at the British newspaper *The Independent* actually applauds her for allowing herself "a teasing chapter on astrology" which "will have shocked the contemporary astronomers who have supplied most of her data" and quips, "It was probably worth doing just for that." Five days later John Gribbin, also at *The Independent* (there may have been complaints), takes Sobel to task for her "uncritical" chapter on astrology and decides she's a "New Age tree-hugger." And Robin McKie at *The Observer* is properly incensed at "the continual, inexcusable references to astrology," quotes from Sobel's comments on Galileo's natal chart, and concludes "This is drivel."

Astrology's foes have been wringing their hands for two hundred years now over the "profitable impostures" foisted on a gullible and unsuspecting public. Astrology has been called trash, rot, balderdash, poppycock, nonsense, hogwash, B.S., rubbish, humbug, bunkum, hocus pocus, and, of course, drivel. Also, folderol, flummery, trumpery, claptrap, twaddle, blather, babble, gibberish, mumbo jumbo, gobbledygook, bosh, flapdoodle, horse feathers, banana oil, tommyrot, malarkey, hokum, hooey, bushwa, baloney, tripe, and hot air. The persistence of the prejudice against astrology is only exceeded by the persistence of astrology itself!

The Merriam-Webster Dictionary defines the word "meme" as "an idea, behavior, style, or usage that spreads from person to person within a culture." American author James Gleick, in his recently published book *The Information*, writes:

Memes can replicate with impressive virulence while leaving swaths of collateral damage – patent medicines and psychic surgery, astrology and satanism, racist myths, superstitions, and (a special case) computer viruses. In a way, these are the most interesting – the memes that thrive to their hosts' detriment, such as the idea that suicide bombers will find their reward in heaven.[207]

In Gleick's opinion, astrology is beyond drivel; it is downright dangerous. Interestingly, Gleick's lumping of astrology with satanism echoes the Roman Catholic catechism of 1994. The condemnation of astrology makes strange bedfellows!

Two recent biographies of Galileo both published at the end of 2010 and reviewed together in *The New York Times Book Review* (an example of synchronicity?) reflect interesting degrees of ambivalence toward their subject's practice of astrology. *Galileo* by J. L. Heilbron, editor of the *Oxford Companion to the History of Modern Science*, the longer and more comprehensive work, brings Galileo out of the closet on the very first page, where the author mentions Galileo's horoscope "which he drew up himself." On page three Heilbron refers to Galileo's "lifetime flirtation" with the art of astrology, which may be an oxymoron: one suspects that a practice that lasts for a lifetime may no longer be dismissed as a flirtation. The purpose of the word, of course, is to trivialize astrology. But Heilbron's book does include a healthy dose of accurate information about Galileo's practice of the art, and that is quite unusual.

Not so David Wootton's *Galileo: Watcher of the Skies*. Wootton who, according to Owen Gingerich's review,[208] has written repeatedly about atheism, makes a few passing references to

Galileo's astrology. More than once, he hastens to inform the reader that "to the lay mind there was no difference between astronomy and astrology." As if a regrettable confusion of the two terms was attributable to the ignorance of the uneducated, whereas in fact at the dawn of the seventeenth century astronomy and astrology were still one science, and even among the educated, the terms were interchangeable.[209]

There is more than a little elitism in the work of some of these modern skeptics, atheists, and proponents of the random universe. In Pulitzer Prize winning author Marilynne Robinson's little book *Absence of Mind*, a scholarly and articulate defense of religion from the attacks of modern so-called rationalists, I was struck by a trait she has in common with my father: their mutual respect for human nature, and for all that humans have created over the centuries. The debunkers of religion and astrology see ordinary humans as gullible, fearful, and deluded, and they see priests and astrologers alike as taking advantage of the gullibility and ignorance of the uneducated masses. In what Robinson calls "a hermeneutics of condescension," these pseudo-rationalists assume their own superiority. Since science taught for centuries that the Earth was flat and did not move, a little humility might be called for.

It has always seemed to me that the focus on gullibility on the part of the modern skeptics may well be a case of projection: the fear of their *own* gullibility, and the professional humiliation that any cracks in their rationalist armor might entail. The fear (quite rational in this case) is that any openness to a subject such as astrology might subject them to the sort of ridicule they dish out: that astrology, like some dread disease, might be contagious. If there is any validity at all to the astrological hypothesis, then the whole materialist world view is called into doubt. It's Copernicus all over again.

It also seems to me that the fear of astrology – for under the logic is aversion, and under the aversion is fear – is a fear of complexity. According to the reductionist philosophy, all is accident, all is random. The monism in rational materialism is a flight from complexity. In some ways, modern science is a kind of fundamentalist religion, and to be open to astrology is tantamount to heresy and is punishable by excommunication. The tenets of this religion are that nothing that can't be weighed and measured exists, that all is random and chance, that the universe is composed of dead matter and is devoid of meaning. To see meaning where none exists is as bad as attributing human thoughts and feelings to animals, in a kind of cosmic anthropomorphism.

Robinson notes that these pseudo-rationalists proceed by narrowing definitions and ignoring the evidence:

I have come to the conclusion that the random, the accidental, have a strong attraction for many writers because they simplify by delimiting. Why is there something rather than nothing? Accident. Accident narrows the range of appropriate strategies of interpretation, while intention very much broadens it. Accident closes on itself, while intention implies that, in and beyond any particular fact or circumstance, there is vastly more to be understood. . ..

One would think that the inadequacy of any model to deal with the complexity of its subject would make its proponents a bit tentative, but in fact the tendency of the kind of thought I wish to draw attention to is to deny the reality of phenomena it cannot accommodate, or to scold them for their irksome, atavistic persistence.[210]

As we have seen, a favorite argument against astrology is that it is based on a magical world view in which no distinction is made between the physical, measurable universe and the psychological realm: the view that everything in the universe – the macrocosm – is reflected or has correspondence with everything in the human being – the microcosm. But this so-called magical world view, in which all things are alive and connected by a body of correspondences and meanings, is not that far from the modern conception of Gaia, the belief that the Earth is a living organism rather than a ball of dead matter. In a sense, the ancient astrologers were among the first ecologists.

There is something to be said for the magical world view. The so-called "primitive" inhabitants of this continent, who believed everything was alive and full of meaning, took much better care of the land. Deconstruction involves the draining of meaning and purpose from the outer world, which is reduced to pure object, a resource to be exploited.

Campion writes that "the creation of meaning is a fundamental, essential human activity. We cannot function without it":

In Plato's moral universe the entire cosmos is "alive" in the sense that it is an extension of God's mind. And when Stephen Hawking famously said that his goal was to know the mind of God he was part of a noble tradition. The distinction is that in Plato's cosmology, which remained the dominant model in Europe until the seventeenth century, the universe was held together by the world soul, the Latin *anima mundi*, as a living organizing principle. In the disenchanted, de-animated mythology of modern astronomy, the world soul has been replaced by the unified field theory, the supposed principle which underpins the known nuclear and electromagnetic forces and reconciles classical and quantum physics.[211]

To dismiss astrology out of hand is too simple. At the end of the day, I suspect that the universe is neither one hundred percent random nor one hundred percent meaningful. The truth may lie somewhere between these extremes. The truth, as in mathematics, is both complex and simple: in the mathematician's favorite word, elegant. Nuance is called for, and above all, attention.

Gauquelin, our modern Kepler, first committed professional suicide by finding something in astrology that was validated by rigorous testing. And when his forty years of pioneering and painstaking work brought him nothing but ostracism and rejection, he took his own life: a martyr to so-called rational materialism's abdication of its own scientific method.

In the mid-twentieth century many astrologers focused their attention on the horoscope as a path to self-knowledge in the Rudhyar tradition rather than on the precise prediction of future events. Indeed, the emphasis on character analysis rather than prediction has enabled many Christians to reconcile astrology with their faith. My college roommate, now a Benedictine nun, tells me:

I have never had any conflict between my Catholic faith and astrology. The Church does prohibit using divination; but there is far more to astrology than that. I have no desire to probe my future. What I appreciate about astrology is that, given a good practitioner, one can learn about oneself in a much deeper way; and perhaps learn a bit about others, too.[212]

But the recent revival of very ancient texts has unearthed many that are full of detailed instructions for prediction, and there are some modern astrologers who are astonishingly accurate at prediction.

One of the pioneers of this school is American astrologer and medievalist Robert Zoller, who predicted the attack of September 11 in a series of forecasts, starting several years ahead. In the September 2000 issue of his newsletter *Nuntius*, he wrote:

I again draw attention to the increasing threat of Islamic terrorism and that it will be felt on the U.S. mainland. The greatest period of danger is in September 2001. . . .. I am looking at Islamic terrorism that is rooted in Islamic fundamentalism and directed at everyday citizens going about their daily business in our own cities. The destruction and loss of life will shock us all. I repeat my warning now *for the third time* [emphasis mine] that unless the U.S. remains vigilant it will be caught unprepared and we will be rocked to our very core. It will be an act of war but unlike any other in our history. Our culture and way of life and lives are at risk.[213]

Zoller also predicted the stock market decline. Benson Bobrick comments,

It would appear that a lowly but capable astrological scholar poring over a handful of tell-tale charts – using the mundane predictive methods [of traditional medieval astrology] – knew more about what was likely to happen that day than all the intelligence experts in the FBI and CIA combined.[214]

Scientists have deliberately removed their attention from astrology. If they actually paid attention to it – if they could find the courage to do this, in the face of the ridicule of their colleagues – both science and astrology would change. For, as McGilchrist and others have noted, "the nature of attention alters what it finds."[215]

NOTES

1. The source for the Margaret Mead quote about men suffering from retirement is "Growing Old in America: An Interview with Margaret Mead," in *Family Circle Magazine*, July 26, 1977.

2. The quotation from Max Mason is in Warren Weaver, "Max Mason: 1877-1961: A Biographical Memoir," 1961, The National Academy of Sciences.

3. Unless otherwise noted, the quotations in this chapter are all taken from my father's autobiography, *Scene of Change*: *A Lifetime in American Science.* New York: Charles Scribner's Sons, 1970.

4. The epigraph is taken from Warren Weaver, "Confessions of a Scientist-Humanist," *Saturday Review,* May 28, 1966, page 15.

5. Unless otherwise noted, the quotations in this chapter are taken from *Scene of Change*.

6. In 1965 Warren Weaver was awarded the first Arches of Science medal for outstanding contributions to the public understanding of science, and in the same year, he received UNESCO's Kalinga prize for distinguished contribution to the popular understanding of science.

7. The memorandum, entitled simply "Translation," was reproduced in Locke, William. N. and Booth, A. Donald, eds., *Machine Translation of Languages: Fourteen Essays*. Cambridge, Mass.: Technology Press of the Massachusetts Institute of Technology, 1955, pp. 15-23 and/or New York: Wiley, 1955.

8. Silberman, Steve, "Talking to Strangers," in *Wired*, May 2000, pp. 225 ff.

9. As for who invented the term "molecular biology," my father would no doubt insist on sharing the honors with the British crystallographer William Astbury, who was one of the first to use the term. Neither he nor Dad could remember who actually "coined" it, but Dad's use of it in the annual report of the Rockefeller Foundation for 1938 may well have been the first published use.

10. The quotation from Leonardo da Vinci is on the first page of my father's student notebook at the University of Wisconsin.

11. Warren Weaver, "Chance," *The Wisconsin Engineer*, 1916, pp. 52-3.

12. Warren Weaver, "The Reign of Probability," *The Scientific Monthly*, Vol. XXXI, November, 1930, pp. 464-6.

13. Warren Weaver, "Probability, Rarity, Interest, and Surprise," *The Scientific Monthly*, Vol.

LXVII, December, 1948, p. 390.

14. Warren Weaver, "Probability," *The Scientific American*, October, 1950, Vol. 183, No. 4, p. 46.

15. Warren Weaver, *Lady Luck: The Theory of Probability*. New York: Dover Publications, Inc., 1982.

16. Isaac Asimov, in a letter of April 4, 1963 to B. F. Kingsbury of Educational Services, the textbook imprint of Doubleday.

17. Warren Weaver, a letter to the editor of *The Scientific Monthly*, Vol. LXX, No. 2, February, 1950.

18. Much of this chapter is taken from my book *The Awakener: A Memoir of Kerouac and the Fifties* (City Lights, 2009).

19. The epigraph is from *Science and Imagination: Selected Papers of Warren Weaver*, New York and London: Basic Books, Inc., 1967, page 88.

20. Warren Weaver, "Peace of Mind," in *Saturday Review*, December 11, 1954, p. 11.

21. Warren Weaver, "Science and Faith," in *The Christian Century*, Vol. LXXII, Number 1, January 5, 1955, pp. 10-13.

22. Warren Weaver, "Can a Scientist Believe in God?" *Look*, April 5, 1955.

23. Originally published in *Look* for April 5, 1955 and then in condensed form in *Readers Digest* for July 1955, "Can a Scientist Believe in God?" was translated into eight foreign editions of *Reader's Digest*: German, Swedish, Norwegian, Danish, Japanese, Italian, Spanish, and Portuguese. It was published as a chapter in Leo Rosten's *Religions of America* which was translated into German as *Zu Gott Furen Vielen Wege*, and it was published in Indonesian by the U.S. Information Service.

24. "Ant Illumination," and much of this chapter, is taken from *The Awakener*.

25. Warren Weaver, "In Pursuit of Lewis Carroll," *The Library Chronicle of the University of Texas*, November 1970 (pages unnumbered).

26. Dad's version of the classic turtles story was supplied to me by the late Dennis Flanagan, who was editor of the *Scientific American* for thirty-seven years. Flanagan sent it to me in a letter dated February 14, 1989. About my father he said, "He was certainly one of the very best people

I ever met."

27. Warren Weaver, *Science and Imagination*, p. 122.

28. Warren Weaver to Selwyn Brant, June 4, 1968.

29. The epigraph is from Jung's *Modern Man in Search of a Soul*.

30. Ihsan Taylor, in a thumbnail review of Benson Bobrick's *The Fated Sky*, January 14, 2007.

31. Later I learned that in the case of a doubtful birth time, or no time at all, it makes more sense to cast a chart for noon on the day of birth, since a noon chart cannot be more than twelve hours off the actual (unknown) time.

32. Quoted in the *Larousse Encyclopedia of Astrology*, McGraw-Hill Book Company, 1977.

33. The University of Texas paid my father $65,000 for his collection. Of course, it would be worth much more than that today.

34. See Jung, *Memories, Dreams, Reflections*, New York: Vintage Books, 1963, pp. 27-28, where Jung describes his "downright fear of the mathematics class," and cites this very proposition as the most exasperating part of it. Had my father read this book, or is this an example of synchronicity? I'll never know!

35. The epigraph is in a letter quoted in *Science and Imagination*, p. 284.

36. An expanded and updated edition of *The Case for Astrology* by John Anthony West was published by Viking in 1991.

37. For a detailed account of Gauquelin's work, see John West, *The Case for Astrology*, New York: Viking Arkana, 1991; for an account of the controversy it aroused, see Suitbert Ertel and Kenneth Irving, *The Tenacious Mars Effect*, London: The Urania Trust, 1996.

38. Warren Weaver, "Science and Complexity," in *American Scientist*, Autumn 1948, Vol. 36, o. 4, pp. 542-3.

39. Literally, "beneath the moon;" hence, of or pertaining to this world; terrestrial, mundane, earthly.

40. Vernon Clark, "An Investigation of the Validity and Reliability of the Astrological Technique," *In Search*, Winter 1960, Vol. II No. 4, p. 48. The late great astrologer Charles Emerson kindly provided me with copies of his letters from Vernon Clark about the experiments

they collaborated on from 1958 to 1966.

41. Warren Weaver, "Science and Older Wisdom," *Academy Reporter* (newsletter of the Academy of Religion and Mental Health), Vol. 5, No. 7, October, 1960, pp. 1 and 4.

42. See Brown, Frank A., "The rhythmic nature of animals and plants," *NorthWestern University Tri-Quarterly*, Fall, 1958; and Brown, Frank A. *et al*, "Monthly cycles in an organism in constant conditions during 1956 and 1957," *Proceedings of the National Academy of Sciences*, April, 1958.

43. See Nelson, John H., "The cyclic elements of radio weather forecasting," *Journal of Cycle Research*, January, 1961. The "hard" aspects are the opposition, square, and semisquare, angles of 180, 90, and 45 degrees respectively.

44. See Michel Gauquelin, *Cosmic Influences on Human Behavior: The Planetary Factors in Personality*, New York: ASI Publishers Inc., 1973, and Michel and Françoise Gauquelin, *The Planetary Factors in Personality*, Paris: *Labortoire d'Étude des Relations entre Rythmes Cosmiques et Psychologiques, 1976.*

45. Nicholas Campion points out that in the Greek root of the word zodiac, the "zo" can mean not only animals (as in zoology) but life itself. Therefore, zodiac means not only "dial of animals" but also "circle of life." See Campion, *A History of Western Astrology: Volume I: The Ancient World.* London: Continuum Books, 2008, pp. 180-1.

46. The epigraph is from a letter of April 21, 1967 from my father to Hasbrouck Van Vleck, a professor of physics at Harvard.

47. See Max Mason and Warren Weaver, *The Electromagnetic Field*, The University of Chicago Press, 1929, and reprinted by Dover Publications in the 1950s.

48. The epigraph is from Richard Feynman's Nobel Prize address in 1965. He shared the prize in physics with Sin-Itiro Tomonaga and Julian Schwinger.

49. Ruth Hale Oliver, ed., *The Basic Principles of Astrology: A Modern View of the Ancient Science*, American Federation of Astrologers, 1962, p. 5.

50. *Ibid.* p. 7; See also John J. O'Neill, "A Scientist Speaks," *AFA Bulletin*, March 27, 1960, p. 16, and *Horoscope* Magazine, April, 1960.

51. Bart J. Bok, "A Critical Look at Astrology," *The Humanist*, September/October 1975.

52. Niels Bohr, in # J. Gribbin, *In Search of Schrödinger's Cat*, Wildwood House, London,

1984., p. 5.

53. The "leading scientific journal" is *Scientific American*, February, 1976.

54. Dad's footnote reads: "The exponential notation ten to the minus 23rd power is, of course, the simplest and most convenient one, but to most persons it does not indicate or even hint at an understandable interval of time."

55. And in another footnote he writes: "Perhaps this is what Freud had in mind when he said that 'We are born in a state of oneness with the world,' as well as what Joseph Conrad had in mind when, in his essay on 'The Condition of Art,' he referred to 'the latent feeling of fellowship with all creation.' C. G. Jung, in his foreword to *The Secret of the Golden Flower*, said, 'Whatever is born or done in this moment of time has the qualities of this moment of time.'"

56. Richard Feynman *et al*, *The Feynman Lectures on Physics*, California Institute of Technology, 1963; Volume 1, Section 7-7, pp. 7-9.

57. Warren Weaver, "A Scientist Ponders Faith," *Saturday Review*, January 3, 1959.

58. Allen Ginsberg, "Father Death Blues," *First Blues* (audio CD), 2006.

59. *Larousse Encyclopedia of Astrology*, p. 57.

60. *Ibid.*, p. 117.

61. *Ibid.*, p. 123.

62. *Ibid.*, p. 119.

63. Robert Hand, "On the Early Roots of Horoscopic Astrology," in *Chronology of the Astrology of the Middle East and the West by Period* (ARHAT, 1998), p. 21.

64. Nicholas Campion, *A History of Western Astrology: Volume I, The Ancient World*, London: Continuum Books, 2008, p. 4.

65. *Ibid.*, p. 4.

66. *Ibid.*, p. x.

67. Hand, *op. cit.*, p. 24.

68. *Ibid.*, p. 26.

69. *Ibid.,* p. 26.

70. This description of van der Waerden's three phases appears in Hand, *op. cit.,* pp. 26-7.

71. The word "ephemeris" is from the Greek *ephemoros,* "lasting no more than a day;" our word ephemeral is from the same root.

72. Campion, *op. cit.,* p. 83.

73. *Ibid.,* p. 83. For the details of this birth chart, with its fragmentary delineation – the first in recorded history! – Campion cites Sachs, A.J., "Babylonian Horoscopes," in *Journal of Cuneiform Studies* (1952), pp. 54-7, and Rochberg, Francesca, *Babylonian Horoscopes* (1998), pp. 51, 56-7.

74. The Hermetic texts, attributed to an Egyptian sage who preceded Plato, and even Moses, are actually of multiple authorship. They were once believed to be very ancient, but were probably written in the early centuries A.D. (See Frances Yates, *Giordano Bruno and the Hermetic Tradition*). They were a kind of "varieties of religious experience" of their time. Their cosmology is astrological, but they are more concerned with astral magic than with astrology *per se.*

75. Hand, *op. cit.,* pp. 30-1.

76. Campion, *op. cit.,* pp. 89-90.

77. Gilbert Murray, *Five Stages of Greek Religion,* p. 144, quoted in Koestler, *The Sleepwalkers: A History of Man's Changing Vision of the Universe*: London, Penguin Books, 1959, p. 114.

78. On astrology among the Greeks, see *Larousse Encyclopedia of Astrology,* pp. 124-6; on Hermetic theory, *ibid.,* p. 133.

79. Campion, *op. cit.,* pp. 153-4. The last before the great schism, but fortunately not the last in modern history. See Gauquelin, pp. 227 ff.

80. *Ibid.,* p. 162.

81. Kos is the island Andy and I lived on in the summer of 1961, and the scene of my Ant Illumination. Andy and I actually visited the site of Berosus's school. I remember a huge tree surrounded by a circle of seats.

82. A very good question to which I have not yet encountered a satisfactory answer. I have a

good friend who has an identical twin sister. The two women (born, according to their mother, only five minutes apart) look and sound so much alike that they are frequently taken for one another; yet their characters and personalities could not be more different. All I can come up with is that if the birth time I have been given is accurate, one has 29 degrees of Cancer rising and the other has zero degrees of Leo. All of the other house cusps are different too, but everything else is the same. Makes me crazy!

83. You may also remember that the word "skeptic" comes from the same Greek root as "horoscope:" *skeptesthai*, to view or consider. A skeptic, in its original sense, meant one who doubts, but looks at the evidence. With the decline in religious belief, the word has come to imply one who believes nothing that cannot be explained and measured by science – in other words, a rational materialist.

And the word "consider" comes from the same root as "sidereal:" the Latin *sidus, sideris*, star or constellation. The original meaning of the word was "to observe the stars." Whether we like it or not, astrology is built into our language!

84. See *Larousse*, pp. 224-5.

85. A combination of the Greek god Hermes and the Egyptian god Thoth. According to Wikipedia, they were both gods of writing and magic in their respective cultures, and were patrons of astrology and alchemy.

86. *Deuteronomy* 18:10-12.

87. See *Leviticus* 19:26; *Isaiah* 47:13-14; and *Jeremiah* 10:2.

88. *Genesis,* 1:14.

89. *Luke,* 21:25.

90. *Catechism of the Catholic Church*, Mahwah, New Jersey: Libreria Editrice Vaticana, 1994.

91. Campion, *op. cit.,* Volume 2, p. ix.

92. Campion, *op. cit.,* Volume 1, p. 281.

93. Translation of Light: According to *Larousse*, "a transfer of energy between two planets not within orb of aspect by the agency of a transiting planet that is separating from an aspect to the first while applying to an aspect of the second." What *Larousse* fails to mention (and actually contradicts by a poorly chosen example) is that the planet that "translates" (i.e., carries) the light needs to be one of the faster moving bodies in the solar system, i.e., Moon, Mercury, or Venus. The theory that these techniques may have been Persian is in Hand, *op. cit.*, p. 36.

94. See page 177 for definitions of these techniques.

95. Astrolabe: "from the Greek *astron*, star, and *lambanein*, to take: a compact instrument, said to have been invented by Hipparchus . . . for observing the positions of the heavenly bodies and determining their elevation above the horizon. Astrologers used the astrolabe in order to erect horoscopes before the publication of ephemerides." –*Larousse*, p. 19.

96. Campion, *op. cit.*, Volume II, p. 72.

97. Campion, *op. cit.*, p. 74.

98. On astrology in the fourteenth century, see also H. Darrel Rutkin, "Astrology," in Daston, Lorraine and Park, Katharine, eds., *The Cambridge History of Science, Vol. 3: Early Modern Science*, pp. 541-3.

99. Campion, *op. cit.*, pp. 80-4.

100. Campion, *op. cit.*, p. 103.

101. Campion, *op. cit.*, p. 113.

102. Campion, *op. cit.*, p. 133.

103. The phrase is in Timothy Ferris, *Coming of Age*, quoted in Benson Bobrick, *The Fated Sky*, p. 159.

104. Max Caspar, *Kepler*, 1947 (Dover edition, 1993), p. 16.

105. Kepler, in a letter to the Duke of Wallenstein; quoted in Bobrick, *op, cit.*, p. 167.

106. Kepler, in Caspar, *op. cit.*, p. 59.

107. Kepler, *Tertius interveniens*, quoted in Caspar, *op. cit.*, p. 182.

108. Note that in the elements of astrology that Kepler retained and those he eliminated, he anticipated the work of Gauquelin. The planets and the angles are exactly the same elements of astrology that Gauquelin focuses on three hundred years later.

109. In a letter of April 19, 1610, Kepler wrote Galileo about the semi-sextile, an astrological aspect he had discovered.

110. Kepler objected to precise prediction; yet he did predict the Defenestration of Prague, the dramatic event of 1618 that launched the Thirty Years War.

111. Kepler, in a letter of 1601, quoted in Bobrick, *op. cit.*, p. 166.

112. Quoted in *Larousse*, pp. 164-5.

113. The conception chart was cast for May 16, 1571, at 4:37 am in Weil-der-Stadt (in Caspar, *op. cit.*; origin unknown).

114. Favaro was the first major historian of science to acknowledge Galileo's astrology. See especially *La Libreria di Galileo Galilei*, 1886; cited in H. Darrel Rutkin, *op. cit.*

115. Rutkin, *op. cit.*, pp. 116.

116. *Ibid.*, pp. 117-119.

117. *Ibid.*, pp. 122-3.

118. *Ibid.*, pp. 129-33.

119. For recent exceptions to scholarly avoidance of Galileo's astrology, see Campion, Nicholas and Kollerstrom, Nick. *Galileo's Astrology.* Bristol, England: Culture and Cosmos, 2003; H. Darrel Rutkin, "Galileo Astrologer: Astrology and Mathematical Practice in the Late Sixteenth and Early Seventeenth Centuries", in *Galilaeana: Studi 11*, Firenze: Leo S. Olschki Editore, 2005; and J. L. Heilbron, *Galileo*, Oxford and New York: Oxford University Press, 2010.

120. See Germana Ernst, in Campion, Nicholas and Kollerstrom, *op. cit.*, p. 29.

121. Rutkin, *op. cit.*, p. 109.

122. Peter Machaer, ed., The *Cambridge Companion to Galileo:* Cambridge University Press, 1998.

123. Bacon, in his essay "On the Increase of Knowledge," quoted in Bobrick, *op. cit.*, p. 229.

124. Rutkin, "Astrology," in Daston, Lorraine and Park, Katharine, eds., *The Cambridge History of Science, Vol. 3: Early Modern Science*. Cambridge: Cambridge University Press, 2006, p. 551.
    See also Rutkin, "Various Uses of Horoscopes: Astrological Practices in Early Modern Europe," in *Horoscopes and Public Spheres: Essays on the History of Astrology*, Gunther Oestmann, H. Darrel Rutkin and Kocku von Stuckrad (eds.), Berlin: Walter de Gruyter, 2005, pp. 167-82.

The historically oriented research program Bacon advocates prefigures that of modern historian and astrologer Richard Tarnas (see pp. 234 ff.).

125. On Morin, see *Larousse Encyclopedia of Astrology*, p. 193.

126. On Dryden's astrology, see Bobrick, *op. cit.*, pp. 224-5. See also William Bradford Gardner, "John Dryden's Interest in Judicial Astrology," in *Studies in Philology*, University of North Carolina Press, 1950, pp. 506-521.

127. See Rutkin, "Astrology," in the *Cambridge History of Science*, p. 554. See also Campion, *op. cit.*, p. 165: In accordance with the electional chart chosen by Flamsteed, the Greenwich Observatory was founded August 10, 1675, at 10:03:14.

128. The cover of Poor Richard's Almanack for 1739 appears on Wikipedia at http://en.wikipedia.org/wiki/Poor_Richard_27s_Almanack.

129. Bobrick, *op. cit.*, p. 200.

130. See Nelson, John H., "Planetary Position Effect on Short Wave Signal Quality." in *Electrical Engineering 71*, no. 5 (May 1952), pp. 421-24; and "Shortwave Radio Propagation Correlation with Planetary Positions," in *RCA Review 12* (March 1951), pp. 26-34. See also *Larousse Encyclopedia of Astrology*, pp. 199-200; and West, John Anthony, and Jan Gerhard Toonder, *The Case for Astrology*. New York: Coward-McCann, 1970, pp. 178-80. Nelson's work has been largely ignored by astronomers because his results lend support to the astrological hypothesis.

131. Rutkin, *op. cit.*, p. 556.

132. Isaac Newton, *The Principia*, quoted in Campion, *op. cit.*, p. 175.

133. On astrology's loss of legitimacy in the eighteenth century, see Rutkin, *op. cit.*, pp. 552-3.

134. Rutkin, *op. cit.*, p. 554.

135. *Encyclopaedia Britannica: a Dictionary of Arts and Sciences*, 1768, quoted in Rutkin, *op. cit.*, p. 559.

136. Jacques Casanova, *Histoire de ma vie*, quoted in Campion, *op. cit.*, p. 173.

137. Johann Wolfgang von Goethe, *Autobiography: Truth and Fiction relating to my Life*, Volume 1. BiblioBazaar, 2006, p. 43.

138. *Larousse Encyclopedia of Astrology*, p. 295.

139. A fuller version of this story can be found in Patrick Curry, *A Confusion of Prophets: Victorian and Edwardian Astrology.* London: Collins & Brown, 1992, pp. 18-19.

140. Campion, *op. cit*, p. 216.

141. Patrick Curry, *op. cit.*, p. 13.

142. On CSICOP, see *Ibid.,* pp. 97 and 191-2.

143. *Ibid.*, p. 77.

144. *Ibid.*, p. 79.

145. *Ibid.*, p. 53.

146. *Ibid.*, plate between pp. 96 and 97.

147. *Larousse Encyclopedia of Astrology*, pp. 200-201.

148. *Ibid.,* p. 201.

149. Blavatsky, *Isis Unveiled*, Volume 1, p. 259; quoted in Campion, *op. cit.*, p. 230.

150. According to my lawyer, this statute is still on the books.

151. *Larousse Encyclopedia of Astrology*, p. 21.

152. Mendelson, Edward, ed., *The Complete Works of W.H. Auden*: Prose, Volume II, 1939-1948, 2002.

153. For a definition of precession of the equinoxes, see page 185.

154. Cf. Michael York, *Historical Dictionary of New Age Movements*; quoted in Campion, *op. cit.*, p. 242.

155. *Larousse*, pp. 219-223.

156. *The Spirit in Man, Art, and Literature*; quoted in Campion, *op. cit.*, p. 254.

157. Quoted in Richard Tarnas, *Cosmos and Psyche: Intimations of a New World View*, New

York: Penguin, 2006, p. 103.

158. Jung, September 6, 1947, in *Letters 1*, p. 475.

159. Jung's definition of synchronicity occurs in "Synchr3nicity: An Acausal Connecting Principle," in C. G. Jung and W. Pauli, *The Interpretation of Nature and the Psyche*, New York: Pantheon Books, 1955, p. 36.

160. Jung, *op. cit.*, p. 27.

161. For Jung on the experiment with the birth charts of married couples, see *Ibid*, pp. 60-94.

162. For Campion on the Myers-Briggs Type Indicator, see *op. cit.*, p. 259.

163. *Ibid.*, p. 251 and pp. 179-180.

164. Charles Carter, "The Astrological Lodge of the Theosophical Society," in *In Search*, Spring 1969, p. 13; quoted in Campion, *op. cit.*, p. 326, note 2.

165. Campion, *op. cit.,* p. 248.

166. Daneudhyar, *The Practice of Astrology*, Penguin Books: Baltimore, Maryland, 1970, pp. 25. and 98.

167. See the AFA website at http://www.astrologers.com.

168. See their website at http://www.astrologicalassociation.com.

169. See their website at http://www.isarastrology.com.

170. See their website at http://www.geocosmic.org.

171. See their website at http://www.afan.org.

172. See Kepler College - Wikipedia, the free encyclopedia / en.wikipedia.org/wiki/Kepler_College

173. For more information about Kepler, go to their website at http://www.keplercollege.org.

174. *The Harvard Crimson*, December 12, 1969, http://www.thecrimson.com.

175. "Effect size" is a standard statistical measure. The Mars Effect does not decline with

replication; it remains solid at 22% (chance is 17%).

176. George Abell, Foreword to Gauquelin, *Dreams and Illusions of Astrology*, New York: Prometheus, 1979; quoted in Gauquelin, *Neo-Astrology*, pp. 32-33.

177. See Dennis Rawlins, "sTARBABY," in *Fate*, October 1981, pp. 67-98 (available as a reprint from Llewellyn Publications, P. O. Box 64383, St. Paul, MN 55164-0383); Patrick Curry, "Research on the Mars Effect," in *Zetetic Scholar* 9, Feb/March 1982, pp. 34-52; John Anthony West, *The Case for Astrology*, London: Viking, 1991, pp. 281-3, 287; and Kenneth Irving, "The Mars Effect Controversy," in Suitbert Ertel and Kenneth Irving, *The Tenacious Mars Effect*, London: Urania, 1996, pp. 17-24.

178. Kenneth Irving, *op. cit.*, p. 31.

179. See Suitbert Ertel, "The Tenacious Mars Effect," in Ertel and Irving, *op. cit.*, pp. 10-47.

180. Kenneth Irving, *op. cit.* p. 37. Scholarly debate on this controversy from 1988 to 2000 is published in the *Journal of Scientific Exploration.* See especially Vol. 2, No. 1, pp. 29-51, and 53-82, 1988; Vol. 4, No. 1, pp 85-104, 1990; Vol. 6, No. 3, pp 247-254, 1992; Vol. 7, No. 2, pp. 145-154, 1993; Vol. 7, No. 3, pp. 271-281, 1993; Vol. 11, No. 1, pp. 1-18, and 19-39, 1997. These articles can be accessed free of charge on the JSE website at http://www.scientificexploration.org/journal/

181. Françoise Schneider-Gauquelin, in *Astrological Journal* 33, September-October, 1991.

182. Irving, *op. cit.*, p. 3.

183. Hans Eysenck and David Nias, *Astrology: Science or Superstition?*, 1982.

184. Geoffrey Dean, www.astrology-and-science.com.

185. Hans Eysenck and David Nias, *Astrology: Science or Superstition?* London: Penguin, 1982.

186. Irving, *op. cit.*, p. vi.

187. Michel Gauquelin, *Neo-Astrology: A Copernican Revolution,* London: Arkana, 1991, p. 24.

188. Gauquelin, *The Truth about Astrology*, 1983.

189. For more information on the Gauquelin Research Fund, go to www.astrology-and-science.com.

190. Irving, "Science and Astrology," in *NCGR Research Journal*, Vol. 1, No. 1, Summer 2010, pp. 57-62.

191. Richard Dawkins, *Unweaving the Rainbow*, London: Penguin, 1998; quoted in Campion, *op. cit.*, Volume 2, p. 270.

192. Carl Sagan, quoted in Bobrick, *The Fated Sky*, p.303.

193. Campion, *op. cit.*, p. 270.

194. *Ibid.*, p. 269.

195. *Larousse Encyclopedia of Astrology*, p. 160.

196. Richard Tarnas, *Cosmos and Psyche*, New York: Penguin, 2007, p. 110.

197. Ellen McCaffery, *Astrology: Its History and Influence in the Western World.* New York: Charles Scribner's Sons, 1942, pp. 357-8.

198. Richard Tarnas, *op. cit.*, p. 142.

199. Iain McGilchrist, *The Master and His Emissary: The Divided Brain and the Making of the Western World*, New Haven and London: Yale University Press, 2009, pp. 38 and 133.

200. The law of the excluded middle was one of Aristotle's basic principles. But "in the new system, a statement may have three values: true, false, or indeterminate." –John Pfeiffer, *Scientific American*, quoted in an article by Harry E. Maynard in *The Journal of the International Society for General Semantics*, January, 1994.

201. For McGilchrist on metaphor, see especially *Ibid.*, p. 71.

202. *Ibid.*, p. 244.

203. See Max Weber, *Essays in Sociology*, London: Kegan Paul, Trench, Trubner & Co., 1947, p. 139.

204. Albert Einstein, Preface, Max Planck, "Where is Science Going?" London: Allen & Unwin, 1933, p. 12.

205. McGilchrist, *op. cit.*, p. 460.

206. Dava Sobel, *The Planets*, New York: Viking, 2005, pp. 158-9.

207. James Gleick, *The Information*, New York: Pantheon Books, 2011, p. 315.

208. Owen Gingerich, "Starry Messenger," *New York Times Book Review*, December 24, 2010.

209. The *New York Times Book Review* is a very accurate reflection of the philosophical assumptions of mainstream modern academics. For example, in "The Stars Can't Help It," Dick Teresi's review of Benson Bobrick's *The Fated Sky* (February 5, 2006), Teresi defines astrology as "the belief that human lives are ruled by the stars and planets."
    In a letter to the editor I wrote, "Some Americans may believe this, but no serious astrologer does. Modern astrologers believe, as Jung did, that a study of the positions of the planets at a person's birth can yield considerable insight into their personality and character. . .. The idea that our lives are ruled by the planets is outmoded and smacks of fortune telling. By providing useful information, astrology can actually enhance our free will rather than overrule it."
    Of course, the *Times* did not print my letter, since it violates their beliefs about astrology.

210. Marilynne Robinson, *Absence of Mind*, New Haven and London: Yale University Press, 2010, pp. 71-2.

211. Campion, *op. cit.*, Volume 1, pp. 5 and 166.

212. My old friend prefers to remain anonymous.

213. Robert Zoller, quoted in Bobrick, *The Fated Sky*, p. 299.

214. *Ibid.*, pp. 299-300.

215. McGilchrist, *op. cit.*, p. 393.

BIBLIOGRAPHY

Bobrick, Benson. *The Fated Sky: Astrology in History*. New York: Simon & Schuster, 2005.

Brau, Jean-Louis, Weaver, Helen, and Edmands, Allan. *Larousse Encyclopedia of Astrology*. New York: McGraw-Hill Book Company, 1980.

Campion, Nicholas. *A History of Western Astrology: Volume I: The Ancient World*. London: Continuum Books, 2008.

Campion, Nicholas. *A History of Western Astrology: Volume II: The Medieval and Modern Worlds*. London: Continuum Books, 2009.

Campion, Nicholas and Kollerstrom, Nick. *Galileo's Astrology*. Bristol, England: Culture and Cosmos, 2003.

Caspar, Max. *Kepler*. Translated and edited by C. Doris Hellman. New York: Dover Publications, Inc., 1993. (Originally published in 1959 by Abelard-Schuman, Ltd., in London.)

Curry, Patrick. *A Confusion of Prophets: Victorian and Edwardian Astrology*. London: Collins & Brown, 1992.

Eysenck, H. J. and Nias, D. K. B. *Astrology: Science or Superstition?* New York: St. Martin's Press, 1982.

Gauquelin, Michel. *The Cosmic Clocks: from Astrology to a Modern Science*. New York: Henry Regnery Company, 1967.

Gauquelin, Michel. *Cosmic Influences on Human Behavior: The Planetary Factors in Personality*. New York: ASI Publishers Inc., 1973.

Gauquelin, Michel. *Neo-Astrology: A Copernican Revolution*. New York and London: Arkana, 1991.

Gauquelin, Michel. *The Scientific Basis of Astrology: Myth or Reality*. New York: Stein and Day, 1970.

Gauquelin, Michel. *Written in the Stars: The Proven Link Between Astrology and Destiny*. Wellingborough: The Aquarian Press, 1988.

Hand, Robert. *Chronology of the Astrology of the Middle East and the West by Period*. Reston,

Virginia: ARHAT, 1998.

Hand, Robert. *Essays on Astrology*. Rockport, Mass.: Para Research, 1982.

Hand, Robert. *Horoscope Symbols*. Rockport, Mass.: Para Research, 1981.

Heilbron, J. L. *Galileo*. Oxford: Oxford University Press, 2010.

Hone, Margaret E. *The Modern Text-Book of Astrology*. London: L. N. Fowler & Co. Ltd., 1951.

Irving, Kenneth and Ertel, Suitbert. *The Tenacious Mars Effect*. London: The Urania Trust, 1996.

Jung, Carl G. *Synchronicity: An Acausal Connecting Principle*. In Jung, C. G. and Pauli, W., *The Interpretation of Nature and the Psyche*. New York: Pantheon Books. Inc., 1955.

Koestler, Arthur. *The Sleepwalkers: A History of Man's Changing Vision of the Universe*. London: Arkana / Penguin Books, 1989. (Originally published by Hutchinson in 1959.)

Lilly, William. *An Introduction to Astrology*. Hollywood: Newcastle Publishing Company, 1972. (Originally published in 1947.)

Locke, William Nash and Booth, Andrew Donald. *Machine Translation of Languages*. New York: John Wiley Sons, Inc. and London: Chapman & Hall, Ltd., 1955.

McCaffery, Ellen. *Graphic Astrology: The Astrological Home Study Course*. Richmond, Virginia: Macoy Publishing Company, 1931.

McGilchrist, Iain. *The Master and His Emissary: The Divided Brain and the Making of the Modern World*. New Haven and London: Yale University Press, 2009.

Robinson, Marilynne. *Absence of Mind: The Dispelling of Inwardness from the Modern Myth of the Self*. New Haven and London: Yale University Press, 2010.

Rudhyar, Dane. *The Astrology of Personality: A Re-formulation of Astrological Concepts and Ideals, in Terms of Contemporary Psychology and Philosophy*. Garden City, New York: Doubleday, 1970.

Rudhyar, Dane. *The Lunation Cycle: A Key to the Understanding of Personality*. St. Paul, Minnesota: Llewellyn Publications, 1967.

Rudhyar, Dane. *The Practice of Astrology: As a Technique in Human Understanding*. Baltimore, Maryland: Penguin Books, 1968.

Rutkin, H. Darrel. "Astrology," in Daston, Lorraine and Park, Katharine, eds., *The Cambridge History of Science, Vol. 3: Early Modern Science*. Cambridge: Cambridge University Press, 2006, pp. 541-61.

Rutkin, H. Darrel. *Astrology, Natural Philosophy, and the History of Science, c. 1250-1700: Studies Toward an Interpretation of Giovanni Pico della Mirandola's Disputationes Adversus Astrologiam Divinatricem.* Ph.D. dissertation, Indiana University, Bloomington, 2002.

Rutkin, H. Darrel. *Galileo Astrologer: Astrology and Mathematical Practice in the Late Sixteenth and Early Seventeenth Centuries.* Firenze: Leo S. Olschki Editore, 2005.

Rutkin, H. Darrel. *Various Uses of Horoscopes: Astrological Practices in Early Modern Europe.* In Oestmann, Gunther, Rutkin, H. Darrel, and von Stuckrad, Kocku, eds., *Horoscopes and Public Spheres: Essays on the History of Astrology*. Berlin: Walter de Gruyter, 2005, pp. 167-82.

Shannon, Claude E. and Weaver, Warren. *The Mathematical Theory of Communication.* Urbana and Chicago: The University of Illinois Press, 1949.

Silberman, Steve. "Talking to Strangers," in *Wired*, May 2000, pp. 225 ff.

Sobel, Dava. *Galileo's Daughter: A Historical Memoir of Science, Faith, and Love.* New York: Penguin Books, 1999.

Sobel, Dava. *The Planets.* New York: Viking, 2005.

Swerdlow, N. M. *Galileo's Horoscopes.* Chicago: Science History Publications Ltd., 2004.

Tarnas, Richard. *Cosmos and Psyche: Intimations of a New World View.* New York: Penguin Plume Book, 2007.

Tarnas, Richard. *The Passion of the Western Mind: Understanding the Ideas that have Shaped our World View.* London: Random House Pimlico, 1996.

Tarnas, Richard. *Prometheus the Awakener: An Essay on the Archetypal Meaning of the Planet Uranus.* Woodstock, Connecticut: Spring Publications, 1995.

Weaver, Warren. *Alice in Many Tongues: The translations of Alice in Wonderland.* Madison: The University of Wisconsin Press, 1964.

Weaver, Warren. *Lady Luck: The Theory of Probability.* New York: Dover Publications, Inc., 1982.

Weaver, Warren. *Scene of Change: A Lifetime in American Science.* New York: Charles Scribner's Sons, 1970.

Weaver, Warren. *Science and Imagination: Selected Papers of Warren Weaver.* New York and London: Basic Books, Inc., 1967.

West, John Anthony and Toonder, Jan Gerhard. *The Case for Astrology.* New York: Coward-McCann, Inc., 1970.

West, John Anthony. *The Case for Astrology.* London: Viking Arkana, 1991.

Wootton, David. *Galileo: Watcher of the Skies.* New Haven and London: Yale University Press, 2010.

Yates, Frances A. *Giordano Bruno and the Hermetic Tradition.* Chicago and London: University of Chicago Press, 1964.

ACKNOWLEDGMENTS

Behind every book there's a team: those who have inspired it, those who have never stopped believing in it, those without whom it would never have come to be.

I give thanks to all astrologers, living and dead, but especially to: Johannes Kepler, the first of the modern reformers; Michel Gauquelin, victim of our modern ignorance of spirit, whose courage and persistence I'll never forget; and Robert Hand, surely our greatest living astrologer, who is almost singlehandedly responsible for rescuing ancient texts from obscurity. You are my guiding stars and I thank you from the bottom of my heart.

As I do scholars Richard Tarnas, Darrel Rutkin, and John Anthony West, who dare to take astrology seriously in our skeptical age.

I am grateful to Vernon Clark, whose early tests of astrology helped to open my father's mind, and Al Morrison, without whose kindness in sending me the answers to the Vernon Clark test this story could not have its happy ending.

Also to Steve Silberman, whose article in *Wired* magazine introduced me to my father's famous memorandum about machine translation.

And of course, deep gratitude to my dear father who felt we had a book before I knew it myself; and to dear friends who refused to let me use illness as an excuse not to finish it. They reminded me that my beloved Proust finished his book in bed, and so could I.

And my heartfelt thanks go out to my team of free editors: Sally Weaver, Marcia Newfield, Cynthia Poten, Clarisse Zielke, and Mother Felicitas Curti, OSB, a Benedictine nun who has no trouble reconciling astrology with Christianity. Each of them has helped make it a better book.

My heart breaks that my great friend and editor Miriam Berg did not live to see me finish it. All my love to her.

Finally, my gratitude to Maria Kay Simms and Tom Canfield for their generous support over the years.

Not all astrologers are lucky enough to have the support of their families. (That's why I wrote the book!) One of my early readers wept when she came to the part where my father had what Jung called "a change of attitude." I thank her for her tears.

Blessings to all, and may we live to see the queen of sciences come back into her own.

Helen Weaver 2020

If one looks deeply into science, he finds unresolved and apparently unresolvable disagreement among scientists concerning the relationship of scientific thought to reality - and concerning the nature and meaning of reality itself. He finds that the explanations of science have utility, but that they do, in sober fact, not explain. He finds the old external appearance of inevitability completely vanished, for he discovers a charming capriciousness in all the individual events. He finds that logic, so generally supposed to be infallible and unassailable, is, in fact, shaky and incomplete. He finds that the whole concept of objective truth is a will-o'-the-wisp.

Warren Weaver

www.ingramcontent.com/pod-product-compliance
Lightning Source LLC
Chambersburg PA
CBHW060945170426
43197CB00026B/2999